D1196336

National Intelligence and Science

National Intelligence and Science

Beyond the Great Divide in Analysis and Policy

WILHELM AGRELL

and

GREGORY F. TREVERTON

OXFORD
UNIVERSITY PRESS

OXFORD
UNIVERSITY PRESS

Oxford University Press is a department of the University of
Oxford. It furthers the University's objective of excellence in research,
scholarship, and education by publishing worldwide.

Oxford   New York
Auckland   Cape Town   Dar es Salaam   Hong Kong   Karachi
Kuala Lumpur   Madrid   Melbourne   Mexico City   Nairobi
New Delhi   Shanghai   Taipei   Toronto

With offices in
Argentina   Austria   Brazil   Chile   Czech Republic   France   Greece
Guatemala   Hungary   Italy   Japan   Poland   Portugal   Singapore
South Korea   Switzerland   Thailand   Turkey   Ukraine   Vietnam

Oxford is a registered trademark of Oxford University Press
in the UK and certain other countries.

Published in the United States of America by
Oxford University Press
198 Madison Avenue, New York, NY 10016

© Oxford University Press 2015

CIP data is on file at the Library of Congress
ISBN 978–0–19–936086–4

1 3 5 7 9 8 6 4 2
Printed in the United States of America
on acid-free paper

CONTENTS

FIGURES

TABLES

PREFACE

This book is a happy collaboration between a Swede and an American, both longtime students and sometime practitioners of national intelligence. It grows out of more than five years of a project at the Centre for Asymmetric Threat Studies at Sweden's National Defence College. The project, under the rubric of homeland security intelligence, has been sponsored by the Swedish Civil Contingencies Agency (MSB, in its Swedish acronym). We are grateful to the agency and especially to two of its officials, Bengt Sundelius, who began the project, and Mikael Tofvesson, who has been a good colleague throughout. The agency has been generous in its support and also in its openness, letting us follow interesting issues where they took us.

Over its course, the project has produced a dozen issue papers on various challenges facing intelligence, and homeland security intelligence in particular. Those include several first passes at the theme of this book—the interplay of intelligence and science, as both confront ever-more complex and uncertain policy problems. We began the project by producing a book-length survey of the state of research on intelligence and suggestions for the future (Gregory F. Treverton and Wilhelm Agrell, eds., *National Intelligence Systems: Current Research and Future Prospects*, Cambridge University Press, 2009).

For one of us (Treverton), actually writing a book with a colleague—as opposed to editing one—has been a first in a long career. And it has been a pleasure. The collaboration has broadened the view of the book beyond the British and American experiences that usually dominate even comparative studies in intelligence. We have sought to reconcile but not homogenize differences in language, for those differences, we have found, often are the germ of an oblique insight.

We owe the largest debts to our immediate colleagues at the Centre, and then to the dozens of officials and observers, Swedish and international, who have attended our workshops in Stockholm, or talked and corresponded with us there or around the world. Of immediate colleagues, we single out Lars Nicander, director of the Centre and the godfather of the project, Jan Leijonhielm, Gudrun Persson, and Josefine dos Santos, as well as Magnus Ranstorp, who prodded us from the angle of the terrorist threat. We also owe debts to Catarina Jönsson, Björn Brenner, Linus Gustafsson, and Linnéa Arnevall, who at various times made the trains run on time and much, much more. We also thank our reviewers at Oxford, one of whom, Sir David Omand, has also been an adviser throughout the project. All these good people should be implicated in whatever is good in this book but spared identification with any shortcomings, which are ours alone.

Santa Monica and Lund, October 2013

Gregory F. Treverton Wilhelm Agrell

National Intelligence and Science

1

Introduction

The Odd Twins of Uncertainty

On December 7, 2012, seven of the Nobel laureates of the year gathered in the Bernadotte library at the Royal Castle in Stockholm for the recording of the annual TV show *Nobel Minds*.[1] Zeinab Badawi, a British journalist, started the discussion by asking the distinguished scientists for their views about the role of science in contemporary societies. Did science matter only if it delivered practical applications? Most of the scientists, while acknowledging the tremendous role of science in social and economic progress, answered no. Science, as they saw it, was mainly curiosity driven, and if the outcome happened to be useful, so much the better. Commercial interests and prospects for commercialization, while increasingly powerful drivers in science policy and research financing, were nevertheless not on the scientists' minds. Rather, the principal force for them was the prospect of investigating interesting problems. There was an inevitable gap, though, between the long-term perspectives of science and the short-term perspectives of politics. That long-term perspective is imperative for science, but it is also important for politicians who have to think about the next election; they would be better off with more scientific competence and ability to comprehend scientific thinking.

However, what then about public expectations of science to deliver definite answers? Medicine laureate Sir John B. Gurdon compared the roles of science and weather forecasting, the latter often criticized for being in error. Even so, we are better off with sometimes inaccurate forecasts than with no forecasts at all; "people should not blame them [weather forecasters] for not getting it exactly right but be grateful for what they do."[2] Zeinab Badawi then raised the

---

[1] The program is available at http://www.nobelprize.org/. The seven participating laureates were Sir John Gurdon (medicine), Serge Haroche (physics), Brian Kobilka (chemistry), Robert Lefkowitz (chemistry), Alvin Roth (economics), David Wineland (physics), and Shinya Yamanaka (medicine).

[2] *Nobel Minds*, SVT Dec. 7, 2012.

question of the responsibility of scientists, considering the conviction of six Italian scientists for their failure to issue a warning prior to the 2009 earthquake that devastated the town of L'Aquila. Serge Haroche (physics) replied that the perception of a zero risk society was an illusion, and that what science is about must be explained better to the public. Shinya Yamanaka (medicine) referred to the situation in Japan, where many laypeople have lost trust in science and scientists after the Fukushima nuclear disaster, due to the flawed risk assessments that preceded it: "All experts kept saying nuclear energy is safe, but it turned out it was not the case." The result was a loss of public trust, and in Yamanaka's opinion the only way to regain it is to be as honest as possible: "to be hundred percent transparent to lay people in words that can be understood."

Nobel Minds 2012 had its fair share of praise for science and the promise of scientific innovations and new frontiers in astrophysics, medicine, or the economy, the latter humbly described as a newcomer with a long way to go toward theoretically understanding what practitioners know (or perhaps think they know). The participants agreed that science had a duty to make the world a better place, by laying the ground for pathbreaking innovations but also by warning against long-term complex risks like global warming. Yet the round table also brought up the darker undercurrent, the pressure for deliverables that simply could not be met, the awkward dilemma between expectations and uncertainty, and the disastrous consequences of loss of trust among the public, who then turn to other providers of pretended knowledge and alternative risk assessments.

Intelligence Minds?

There is no known case of a corresponding round table event involving the world's leading intelligence analysts, and it is hard to imagine one within the foreseeable future, given the reluctance of the vast majority of intelligence organizations around the world to discuss intelligence matters publicly. But for a moment let's imagine Ms. Badawi chairing a similar round table—not *Nobel Minds* but *Intelligence Minds*: What themes could be discussed there? Where would the participants be likely agree or disagree? Probably the participants would all agree not only on a continued need but also on an increased need for intelligence in an age of globalized threats and growing vulnerabilities. They would also be likely to agree on increasing uncertainty, partly due to the complex and fragmented nature of the threats, but also due to a limited and possibly decreasing ability to produce actionable forecasts. Some of the participants would perhaps speak in favor of increased openness on intelligence matters, not in operational details, but to enhance public understanding of and hence acceptance of intelligence as a necessary expenditure and intrusion into the civil liberties of citizens.

The panelists might point to intelligence failures as a cause for declining public and political confidence, and at the same time stress that many of these failures are not just the fault of intelligence services but rather of a broader policymaking context. Intelligence might be wrong, inaccurate, or ambiguous, but failures are by definition a matter of policy, strategy, and operations. The experts around the table would, finally, most likely be able to agree that intelligence had to and could improve, both in terms of international cooperation and enhanced capability to effectively use and analyze information. Some of the experts might be optimistic over future prospects, while others might warn against an over-reliance on intelligence as a response to threats, policy dilemmas, and ever-present uncertainty.

There would certainly have been considerable differences between the two round tables. Much less would have been said around the intelligence table about the role of curiosity, the troublesome interference from financers, and the unpredictability of the output. And nobody would have spoken—as did the scientific opposite number, Shinya Yamanaka—about the need for 100 percent transparency toward laypeople. Still, at some moments a casual viewer could have been confused as to which of the programs he or she had tuned into.

As we discuss further in this book, science and intelligence constitute two remarkably similar and interlinked domains of knowledge production, yet ones that are separated by a deep political, cultural, and epistemological divide. Beyond this separation of academia and intelligence, another and perhaps far more problematic divide becomes increasingly visible, a divide caused by over-stated certainty and loss of trust. Few incidents illustrate this divide more clearly than the failure of Western governments to use intelligence to mobilize domestic and international support for an armed response following the Syrian government's employment of chemical weapons outside Damascus in August 2013. If Fukushima ruined the public trust in scientific assurances, the flawed assessments of Saddam Hussein's weapons of mass destruction in 2002–2003 had a similar effect on the credibility of Western intelligence. If the public was sold "a false parcel of goods"[3] once, it is not likely to accept a similar delivery from the same provider.

Our over-arching goal in this book is to frame intelligence not as a specific activity or separate entity but rather in a broad context of knowledge production. We will thus investigate intelligence not from a traditional intelligence

---

[3] Michael Herman, interviewed by Mike Robinson on "A Failure of Intelligence," in *BBC Panorama* (United Kingdom 2004), 09/07/2004.

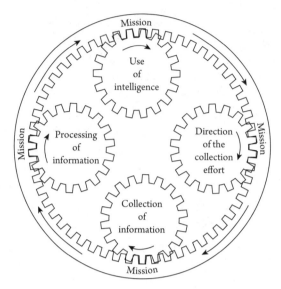

Figure 1.1 Early Depiction of Intelligence Cycle. Source: U.S. Army, 1948.

studies perspective but as one of several modes of knowledge production for action, modes not limited to intelligence but increasingly transcending other fields, including some of those central to the public role of science. Intelligence in this respect is not—irrespective of definitions, organizations, and self-images—a closed state security concern, but a way to define problems, structure data, formulate and test explanations, and manage uncertainty in a social context where decisions have to be made and actions taken. Looking for *intelligence* in this respect leads far beyond traditional boundaries and definitions, but, at the same time, may lead to the discovery of approaches and methods that could in a fundamental way alter the conduct of the existing security intelligence domain.

When we discuss intelligence in this book, we primarily refer to intelligence *analysis,* the process of evaluation, interpretation, and assessment, constituting the final stages in the classical (in)famous intelligence cycle. More on the cycle later. Sir David Omand has drawn attention to what is so far the oldest depiction of the cycle we have seen (see Figure 1.1).

Major intelligence achievements, like the British breaking of the German Enigma code in the Second World War, is often described in terms of sheer intellectual virtuosity. But as Michael Herman reminds us, there is another side of modern intelligence—the large-scale factory-like production line that in the Enigma case provided a steady and timely flow of raw intercepts.[4] No matter how bright the *Intelligence Minds* might be, they would achieve nothing

[4] M. Herman, *Intelligence Power in Peace and War* (Cambridge: Cambridge University Press, 1996), p. 283.

outside the realm of vast and increasingly globalized intelligence machineries. It is furthermore increasingly difficult to draw a distinct dividing line between collection and analysis outside organizational charts. In some instances, collection is not only analysis driven but represents an element in an integral process in which collection becomes analysis and separate roles cease to have any meaning.

With its history as a predominantly state-controlled activity, intelligence has developed with a focus on delivering processed raw data and assessments for the benefit of specified users in the security domain of defense, foreign service, and police. Security became increasingly dependent on intelligence and in many fields impossible to attain without it. Within the Cold War security paradigm, intelligence was focused on estimating potential military threats, assessing the risk of armed conflicts, and warning of imminent attacks. Uncertainty was to be reduced through increased reliance on and refinement of technical systems for surveillance, mass storage of data, and rapid dissemination of intelligence products on activities and targets.[5] What might be called "the Soviet Estimate"—the broad analytic effort to understand the Soviet Union—became the largest puzzle ever set, where every increase in resolution was assumed to result in a corresponding decrease in uncertainty.[6]

Intelligence and Science Converging?

Modern science also developed in a context of uncertainty, or rather fields of incomplete and distorted knowledge. Moving from the introverted knowledge production symbolized by the metaphor of the ivory tower, scientific research since the Second World War has emerged as the paramount system for knowledge production for the purpose of innovation, economic growth, policy, and society as a whole. Intelligence, as invented and institutionalized in the 20th century, was distinctly different from this academic world, which was guided by principles of openness and freedom of information incompatible with the core elements of the prevailing intelligence culture—secrecy, compartmentalization, and the need-to-know principle. Academia and intelligence analysis, while interconnected in several ways, thus nevertheless belonged to different

---

[5] Gregory F. Treverton, *Reshaping National Intelligence for an Age of Information* (New York: Cambridge University Press, 2003); Gregory F. Treverton, *Intelligence for an Age of Terror* (New York: Cambridge University Press, 2009); Gregory F. Treverton and Wilhelm Agrell, *National Intelligence Systems* (New York: Cambridge University Press, 2009).

[6] The term was originally used by John Prados in *The Soviet Estimate. U.S. Intelligence Analysis & Russian Military Strength* (New York: Dial Press, 1982). Practically all Western intelligence services struggled with their own variants of this estimate throughout the Cold War.

worlds, divided not only by legal frameworks and professional traditions, but also by values and mutual mistrust. As shown during the Cold War, bridging this divide was far more difficult than further widening it. And yet, university researchers and intelligence analysts tend to work on similar problems, sometimes also with similar materials and methods for data collection.

Since the 1990s, the traditional concept of empirically based distinctly in-house intelligence production has been challenged more and more. The technically based systems were not designed for and thus are unable or inadequate to cope with a wider spectrum of societal risks and emerging threats, not necessarily resembling the classical military challenges of the bipolar security system. Furthermore, the focus on data left intelligence with a limited capability to cope with complex risk and threat-level assessments. The inability of post–Cold War intelligence to deliver accurate, timely, and verifiable assessments has been displayed time and again. Why intelligence fails to estimate risks and frame uncertainties, to produce the kind of actionable knowledge demanded by policymakers, state institutions, the international community and not least the public, has become a key issue in the literature on the challenges facing intelligence.[7] While mostly elaborated from an Anglo-Saxon perspective, the phenomenon as such appears to be universal.[8] Across the Middle East, the Arab Spring 2011 obviously took most regimes and their respective intelligence and internal security apparatuses by surprise, as was the case with international partners and observers.[9]

The point of departure for this book is the observation of two simultaneous, and possibly converging, trends with a common denominator in the rise of complex societal risks with a high degree of uncertainty. While the Soviet estimate appeared to become less uncertain with increased resolution, the opposite seems to be the case with complex risks: as more data becomes available, the more the web of interlinked uncertainties on possible connections and interpretations tends to grow.

---

[7] For some of the key scholarly works on this theme, see Richard K. Betts, *Enemies of Intelligence: Knowledge and Power in American National Security* (New York: Columbia University Press, 2007); Phillip H. J. Davies, "Intelligence Culture and Intelligence Failure in Britain and the United States," *Cambridge Review of International Affairs* 17, no. 3 (2004): 495–520; Robert Jervis, *Why Intelligence Fails: Lessons from the Iranian Revolution and the Iraq War* (Ithaca, NY: Cornell University Press, 2010).

[8] Several international examples of factors limiting the performance of intelligence organizations can be found in Phillip H. J. Davies and Kristian C. Gustafson, *Intelligence Elsewhere: Spies and Espionage Outside the Anglosphere* (Washington, DC: Georgetown University Press, 2013).

[9] For the Israeli inability to predict the upheaval, see Eyal Pascovich, "Intelligence Assessment Regarding Social Developments: The Israeli Experience," *International Journal of Intelligence and CounterIntelligence* 26, no. 1 (2013): 84–114.

The first trend is an increasing demand for analytic skill, lessons learned, and the long awaited transformation of intelligence into a more "scientific" form.[10] The need was already recognized early in the Cold War, but at that time the thought was that intelligence analysis should develop into a traditional positivist social science discipline.[11] However, half a century later, the epistemological basis for intelligence assessments tends to consist of a rather unsophisticated mixture of common sense, brainstorming, and established practice within a closed profession—and as such not comprehensible to outsiders.[12]

Faced with an increasing complexity in the targets and a surge in demand for detailed, timely, and accurate assessments of a wide range of external and/or internal threats to societies, intelligence analysis is under enormous pressure to transform from a state of proto-science in order to deliver. The potentially devastating social consequences of performance failures underscore the demand for methods to handle uncertainty and to validate assessments. In order to achieve this, intelligence structures are forced to move away from the inherited intelligence culture so as to cooperate between themselves as well as within themselves, a kind of inter-intelligence and trans-intelligence similar to the inter-disciplinary and trans-disciplinary approaches of the scientific community, which is experiencing many of the same challenges.

The second trend is the consequences of challenges, performance pressure, and public expectations of policy-relevant science, leading to a rapid transformation of focus of both the scientific research and the academic institutions: from disciplinary research for the sake of knowledge production to an emphasis on multi-disciplinary approaches. Scientific knowledge production in the twentieth century was characterized by disciplinary specialization and fragmentation, in much the same way as intelligence under the Cold War paradigm. With increasing and more complex external demands, emerging research problems with high relevance for the society had to be met with a new structure, one that not only affected how research was organized but also the mode of knowledge production. In fields like urban studies, health, environment, natural resources, and climate change, the dividing line between natural and social sciences has to be crossed, and researchers are forced to draw conclusions and supply scientific advice under increasing

---

[10] See Peter Gill, Stephen Marrin, and Mark Phythian, *Intelligence Theory: Key Questions and Debates* (London: Routledge, 2009); Stephen Marrin, *Improving Intelligence Analysis: Bridging the Gap between Scholarship and Practice* (London: Routledge, 2011).

[11] Sherman Kent, "The Need for an Intelligence Literature," *Studies in Intelligence* 1, no. 1 (1955).

[12] Rob Johnston, "Analytic Culture in the US Intelligence Community: An Ethnographic Study" (Washington, DC: Center for Study of Intelligence, Central Intelligence Agency, 2005).

uncertainty.[13] The scientific failures experienced in the course of this trans-formation are sometime as devastating as the intelligence failures, and their causes have some striking similarities.[14]

There are still only a few examples of more systematic and permanent cross-disciplinary efforts within intelligence. Most cases emerged due either to demand pull or reform push from external forces after more or less obvious performance failures. The obstacles to inter- and trans-disciplinarity in intel-ligence are well known but nevertheless hard to overcome. Legal, institutional, and professional constraints make genuine cooperation among disciplines far more complicated in the intelligence sphere than in scientific research, thereby further strengthening the already strong proto-scientific subcultures. Intelligence inter-disciplinarity is nevertheless emerging by force of necessity, especially in fusion centers dealing with issues such as counterterrorism.[15] However, these centers suffer limitations similar to large-scale interdisciplin-ary research efforts focused on delivery; they tend to represent a lightpost under which the search for the lost key is conducted.

These two ongoing, largely parallel processes in intelligence and scien-tific problem-oriented research constitute a discreet but no less fundamen-tal shift in knowledge production about potential societal risks and complex security threats. That the processes are parallel is to some extent obscured by the traditional divide between the scientific and intelligence domains and the underdeveloped communication between the two. The main hypothesis, elaborated in this book, is that we are witnessing a process in which intelli-gence intentionally and unintentionally is becoming more "scientific," not nec-essarily in the traditional academic disciplinary sense, but resembling more the emerging complex, cross-boundary, and target-oriented research efforts. Simultaneously this problem-oriented inter- and trans-disciplinary research in science is becoming more like intelligence in focusing on risk assessments, probabilities, and warning, and in communicating not only results but also uncertainty with essential stakeholders. What we see is perhaps not so much a bridging of the 20th-century concept of a divide between academia and

---

[13] For a discussion of "preventive science," see Mark Phythian, "Policing Uncertainty: Intelligence, Security and Risk," *Intelligence and National Security* 27, no. 2 (2012): 187–205.

[14] Among the studied cases are the radioactive fallout from the Chernobyl disaster in 1986 and the BSE disease in Britain in the 1980s and 90s. See Angela Liberatore, *The Management of Uncertainty: Learning from Chernobyl* (Amsterdam: Gordon & Breach, 1999), and Tomas Hellström and Merle Jacob, *Policy Uncertainty and Risk: Conceptual Developments and Approaches* (Dordrecht: Kluwer 2001).

[15] For an overview, see Gudrun Persson, *Fusion Centres—Lessons Learned. A Study of Coordination Functions for Intelligence and Security Services* (Stockholm: Center for Asymmetric Threat Studies, Swedish National Defence College, 2013).

intelligence as a development that is moving beyond both these traditional modes of knowledge production, thereby addressing the divide between the knowledge producers and the knowledge users, or to be more precise, between those responsible for assessments under uncertainty and those who have to comprehend, value, and act on those assessments.

Plan of the Book

The main theme of this book is how the concepts of relevance and uncertainty in intelligence and science for policy have developed and converged. Our aim first of all is to describe and frame an ongoing transformation, one that needs no inventors, no advocates, and no architects. The process is under way, driven by the endless demand-pull emerging from the risk society, and the push from interaction of institutions, professions, and public expectations. The challenge ahead is to handle the subsequent consequences of this transformation and to reshape intelligence *and* science in this shifting and interlinked knowledge environment.

The plan of the chapters reflects this moving back and forth between intelligence and science. The next chapter frames the divide between the two, by looking at how wartime exigencies drove intelligence and science together, more of necessity than choice, and how they drifted apart during the Cold War. It asks whether the two are moving back together again, especially as science becomes more entwined with policy. Is the difference between the two thus more one of context, or are they fundamentally different ways of seeing and creating knowledge? Chapter 3 turns to intelligence, especially intelligence *analysis*. It lays out some of the fundamentals in terms of types of problems, categories of needs, and kind of consumers with a particular focus on conveying uncertainty by contrasting the approaches of Clausewitz and Jomini. It is hard to overstate how much the canonical "intelligence cycle," with its primacy to collection, drives intelligence. In that context, the chapter looks at "roads not taken." It wasn't inevitable that intelligence analysis would come to be practiced as it is, especially in the United States.

Chapter 4 turns back to science. Scientific research, unlike intelligence, is not festooned with "failure"; indeed, its successes are more in the public mind. The very essence of science is a process of failed refutations that eventually validate a hypothesis—a process intelligence hasn't the luxury to emulate. Disagreements in science advance knowledge. To be sure, some scientific controversies can be quite the opposite: witness Lysenko and the destruction of Soviet genetics. More recently, as science has become the arbiter of major public policy issues—from acid rain to global warming—its position looks more like that of intelligence.

Medicine and the sciences, both natural and social, are the most obvious points of comparison for intelligence, but Chapter 5 looks further afield— from sociology and anthropology, to archaeology, journalism, and weather forecasting for suggestive parallels in styles of analysis, ways of conveying uncertainty, and data collection that is necessarily partial. Chapter 6 explores those points of contact in more detail, concentrating on medicine and policy analysis. What might be called "policy analysis 2.0" is like intelligence in dealing with enormous uncertainty, with a widened set of stakeholders and increasing transparency surrounding its work. For policy analysis, the change along these dimensions is one of degree; for intelligence, it is closer to a sea-change.

Chapter 7 digs more deeply into these challenges for intelligence. It begins with a post mortem on post mortems: why do investigations of intelligence "failures," especially in the United States, produce so much political heat and so little light in the sense of serious lesson-learning? It then turns to the challenges of transparency and to both the opportunities and pitfalls of what may amount to a paradigm shift, as intelligence moves from conceiving itself as a deliverer of finished products to "customers" and comes to think it is in the business of providing advice—and help—to "clients." If Chapter 7 ends with future opportunities, Chapter 8 takes up an ever-present risk: "politicization." Drawing on older British and Nordic experience as well as more recent American cases, it makes clear that the risk is not that intelligence will be ordered what to say. Rather, the risk for intelligence—indeed for all of knowledge production—is that analysts will be self-deterred from dissent because they know what answer their policy masters want to hear.

The concluding chapter returns to the recurring theme of uncertainty and the dilemmas posed both by the unknown outside-defined domains and paradigms, and the inevitable trade-off between incomplete knowledge and urgent contingencies. Intelligence and science are increasingly called upon to guide or legitimize the management of complex societal risks. Faced with this challenge, new institutional forms are emerging, and old methods are yielding to new approaches—for instance, moving beyond the sequential steps of the canonical intelligence cycle. In a fragmented world of mass data, we are less and less likely to start with requirements or formulated hypotheses and more to begin with bits and pieces that might, but only might, constitute part of an answer to questions not yet posed and perhaps not yet even thinkable.

2

Framing the Divide

In February 1979, the Peace Research Institute in Oslo (PRIO) released a report titled *Intelligence Installations in Norway: Their Number, Location, and Legality*. It was an 80-page typewritten and photocopied research paper prepared by the Norwegian researcher, Nils Petter Gleditsch, and his colleague, Owen Wilkes from the Stockholm International Peace Research Institute (SIPRI).[1] But as the title indicated, the content was far more explosive than the outer appearance and limited circulation would indicate. In the report the two researchers described the structure of technical intelligence collection in Norway, including the location and layout of the various sites, their size, electronic equipment, and possible roles. The purpose of the investigation was to map the intelligence infrastructure to measure the extent and consequences of Norway's integration not only in the North Atlantic Treaty Organization (NATO) but also in US nuclear strategy. These issues had been of considerable domestic concern since the late 1960s, and nuclear weapons were increasingly becoming so throughout Western Europe.[2]

Research on Intelligence, or Intelligence as Research?

For peace research, the nuclear arms race and the deployment of new generations of weapons systems and supporting infrastructure became a central field

---

[1] Nils Petter Gleditsch and Owen Wilkes, *Intelligence Installations in Norway: Their Number, Location, Function, and Legality* (Oslo: Peace Research Institute of Oslo, 1979). A slightly revised version in Norwegian was published as chapter 1 of Owen Wilkes and Nils Petter Gleditsch, *Onkel Sams kaniner. Teknisk etterretning i Norge* (Uncle Sam's Rabbits. Technical Intelligence in Norway) (Oslo: PAX, 1981).

[2] Nils Petter Gleditsch et al., *Norge i atomstrategien. Atompolitikk, alliansepolitikk, basepolitikk* (Norway in the Nuclear Strategy. Nuclear Policy, Alliance Policy, Base Policy) (Oslo: PAX, 1978);

of interest in the 1970s, answering to a wider public concern over the risk of nuclear war.[3] The peace research community shared a sense of urgency with a growing peace movement and saw the role of the researcher as not constrained to the academic system and its disciplinary orientation, but turned outward as a kind of counter-expertise on security issues to what it perceived as a monopoly by the political, military, and arms industry establishments. While not explicitly referring to this context in their report, the authors in follow-up publications listed the major policy issues for which they saw their findings as relevant—for instance, whether the intelligence installations added to or detracted from Norwegian security.[4]

Given this context, the report aroused considerable media attention and soon became the subject of a heated public debate, not so much over the findings as such but the methods employed by the two researchers. By providing detailed descriptions of what were supposed to be secret defense installations, they had, according to the critics, revealed sensitive information and in practice done the job of foreign intelligence organizations by conducting clandestine intelligence collection under the disguise of academic research.

The authors, however, maintained that academic research was precisely what they had done. In the report they had, according to normal scientific standard, described their sources and methods for data collection. Nothing, they maintained, had been acquired by illegal or even ethically doubtful methods. All they had done was to apply basic methods of social science and use open sources and openly available technical descriptions.[5] One of their main sources was in fact the Norwegian telephone directory, where all defense

on the Nordic and Western European peace movements, see *European Peace Movements and the Future of the Western Alliance*, ed. Walter Laqueur and R. E. Hunter (New Brunswick: Transaction Books, 1985).

[3] Beginning in the mid-1970s, the Stockholm International Peace Research Institute (SIPRI) published a large number of books and research reports on the nuclear arms race. See Frank Barnaby and Ronald Huisken, *Arms Uncontrolled* (Cambridge, MA: Harvard University Press, 1975). A more activist approach was taken by other researchers moving closer to the popular movement, as illustrated by a reader that was widely circulated and commented on, *Protest and Survive*, ed. Dan Smith and Edward Palmer Thompson (Harmondsworth: Penguin, 1980).

[4] Owen Wilkes and Nils Petter Gleditsch. "Research on Intelligence or Intelligence as Research," in Egbert Jahn and Yoshikazu Sakamoto, eds. *Elements of World Instability: Armaments, Communication, Food, International Division of Labor, Proceedings of the Eighth International Peace Research Association General Conference* (Frankfurt/New York: Campus, 1981). Norwegian version: *Forskning om etterretning eller etterretning som forskning* (Oslo: PRIO, 1979; expanded version of Wilkes and Gleditsch as ch. 2).

[5] Gleditsch and Wilkes, *Forskning om etterretning eller etterretning som forskning*; Gleditsch and Wilkes, "Research on Intelligence and Intelligence as Research." Incidentally, the backgound of the original report was another ongoing court case regarding alleged revelation of defense secrets. One purpose of the Gleditsch/Wilkes report was to display just how much could be deduced from open sources.

installations appeared. They simply subtracted the installations with an assigned designation and thus ended up with a list of undeclared facilities. The next step was to go to the documents of the Norwegian Union of Government Employees, organizing the civilian personnel in the Norwegian defense sector. There was a specific section of the Union that matched the undeclared facilities. In a complete list of all sections compiled after a Union vote, the number of members in this section had been deleted, but being a verification of a democratic process, the list contained the number of yes and no votes cast for each local Union entity. The figures thus revealed the approximate personnel strength of what turned out as the major Signals intelligence (SIGINT) station at Vadsø in Northern Norway, overlooking the Kola peninsula and the Soviet Murmansk naval base complex.[6]

But was this really research on intelligence or was it intelligence as research? The authors posed this question in the follow-up reports, starting with the observation that the dividing line was hardly distinct; intelligence and research were, they noted, two closely interlinked fields in which assembled data had to be evaluated according to similar criteria. And when it came to analysis, the methods for formulating and testing hypotheses were identical. True enough, an intelligence analyst probably more often than the average researcher had to choose between hypotheses on the basis of incomplete data. But then again, a researcher outside intelligence could also be forced to draw conclusions on the basis of the available but limited data. Not even the closed nature of the process and the limited circulation of the results were, according to Gleditsch and Wilkes, a clear dividing line, since much applied research also had limited transparency. True, there were differences in the moral and legal frameworks for data collection, and also, most researchers would feel uncomfortable if their work were labeled as intelligence.[7]

Yet this was precisely what happened. The Norwegian defense ministry initiated a criminal investigation, and the two researchers were subsequently charged with breach of paragraphs 90 and 91 of the Norwegian Criminal Code, dealing with the protection of national defense. After a long legal process, they were finally convicted by the Norwegian High Court.[8] The researchers were found guilty of disclosing defense secrets, not by employing methods

---

[6] Gleditsch and Wilkes, *Intelligence Installations in Norway: Their Number, Location, Function, and Legality*, p. 11.

[7] Gleditsch and Wilkes, *Forskning om etterretning eller etterretning som forskning*, pp. 1–2. The section on the parallels between intelligence and research appears only in the Norwegian version.

[8] See *Forskning eller spionasje. Rapport om straffesaken i Oslo Byrett i mai 1981* (Research or Espionage. Report on the Criminal Trial in Oslo Town Court in May 1981) (Oslo: PRIO, 1981) for the verdict and a complete record of the proceedings. A shorter version in English appears as *The Oslo Rabbit Trial. A Record of the "National Security Trial" against Owen Wilkes and Nils Petter Gleditsch in the Oslo Town Court, May 1981* (Oslo: PRIO, 1981). See also, *Round Two. The Norwegian Supreme Court vs. Gleditsch & Wilkes,* (Oslo: PRIO, 1982).

of espionage but by the nature of the aggregated output of the data collection. The verdict broadened the whole issue from the subject of the original report to a conflict between the principles of academic freedom and the protection of national security. Leading academics, even those who did not necessarily sympathize with the researchers, nevertheless came out strongly critical against the verdict and the very idea that the employment of open sources could constitute a crime.[9] The two domains seemed, not only in the isolated Norwegian case but throughout the Western world, very far apart, with incompatible values, professional ethos, and perception of their roles in society. Yet the relation had not started that way. Indeed the two worlds had, under external pressure, found each other and developed not only coexistence but also far-reaching cooperation. Bridges had been laid, only to be later burned to the ground.

The Rise of Intelligence and the Mobilization of the Intellectuals

Intelligence, which was "invented" in its modern organized form in the early 20th century shortly before and during the First World War, was then primarily a matter for three well-established professions—the military, policymakers, and diplomats. They all had long traditions in intelligence collection, often described in other words, but the new world, and not least the new kind of warfare, made knowledge of "the others" an essential. No general staff could plan or launch a military campaign without detailed information of the opponent, based on a multitude of sources and transformed into comprehensive knowledge at the disposal of the key decision makers. The former Austro-Hungarian intelligence officer Max Ronge, in his 1930 book on military and industrial espionage, provided one of the first published schemes of a modern all-source intelligence system.[10]

During the First World War, the academic communities still played a marginal role, not only in intelligence but in the war efforts as a whole. None of the major participants in the war had a clear notion of how the potential of scientific

---

[9] "Dommen over Gleditsch Og Wilkes. Fire Kritiske Innlegg" (Oslo: Peace Research Institute of Oslo, 1981). For a discussion of the wider implications of the Norwegian case, see Nils Petter Gleditsch and Einar Høgetveit, "Freedom of Information and National Security. A Comparative Study of Norway and United States," *Journal of Peace Research* 2, no. 1 (1984): 17–45, and Nils Petter Gleditsch, "Freedom of Expression, Freedom of Information, and National Security: The Case of Norway," in Sandra Coliver et al., eds. *Secrecy and Liberty: National Security, Freedom of Expression and Access to Information* (Haag: Nijhoff 1999), pp. 361–388.

[10] Max Ronge, *Kriegs-Und Industrie-Spionage: Zwölf Jahre Kundschaftsdienst* (Wien: Amalthea-Verl, 1930).

knowledge and expertise could be employed in a systematic way. Individual leading academics, such as the head of the prestigious Kaiser-Wilhelm Institute, Fritz Haber, persuaded the German general staff to support his plans to develop chemical weapons, arguing that this would shorten and thereby humanize the war.[11] But on the whole, the field of new technological applications was left to a spontaneous and mostly chaotic flow of inventors and inventions. There was no lack of ideas, but there was a lack of planning, coordination, and management of priorities. The inventors did not know what to invent—or rather in what context they should invent—and the military organizations had, with some notable exceptions, neither the knowledge nor the flexibility to assess and employ new technology in an innovative way.[12] The First World War was dominated by mostly unsuccessful trial-and-error engineering, and academic research, with a few exceptions, had not yet entered the innovation loop.[13] The main reason was a lack of institutions and procedures: the very concept of science-based inventions had first to be invented and proven.

By the end of the 1930s the situation was different. Science and technology, along with research institutes and academic institutions, had emerged as a key factor in an increasing state intervention in the industrialized economies. In the words of U.S. president Franklin D. Roosevelt in 1937, science had become "one of the greatest resources of the nation."[14] All the major powers in the Second World War would employ this resource on a scale and in forms not seen before. Not only were science-based innovations invented but also the whole concept of research policy, planning, and management. This merger changed the interrelationship between the academic world and the society in a profound and irrevocable way.

The war effort not only mobilized practitioners from the hard sciences—the physicists, chemists, and mathematicians—but the emerging social sciences also found their way into the war efforts and thereby further on into postwar science policy.[15] The transformation brought about by World War II can be described as a dual process, one in which academic research left the ivory tower and entered a realm where it was subjected to bureaucratization and

---

[11] On the role of Haber, see Dietrich Stoltzenberg, *Fritz Haber. Chemiker, Nobelpreisträger, Deutscher Jude* (Weinhem: VCH, 1994).

[12] Jonathan Shimshoni, "Technology, Military Advantage, and World War I: A Case for Military Entrepreneurship," *International Security* 15, no. 3 (1990): 187–215.

[13] Guy Hartcup, *The War of Invention: Scientific Developments, 1914–18* (London: Brassey's Defence Publishers 1988).

[14] Jean Jacques Salomon, *Science and Politics* (Cambridge, MA: MIT Press, 1973), p. 31.

[15] For the role of sociology, see Peter Buck, "Adjusting to Military Life: The Social Sciences Go to War 1941–1950," in *Military Enterprise and Technological Change: Perspectives on the American Experience* (Cambridge, MA: MIT Press, 1985).

militarization, while the military organizations, on the other hand, became dependent on and compelled to adapt to scientific research as a key strategic resource in warfare.[16]

Intelligence was, or rather developed to become, one of the fields subjected to this dual process, although with some significant particularities. The most obvious similarity with the overall pattern of war-related research appeared in the development of the technical means for intelligence collection. Radar, until the atomic bomb, generally ranked as the most important technical innovation put in large-scale use during the war, was both an integral part of aerial and naval weapons systems and a technique for intelligence collection (and deception in the form of counter-measures).[17] But intelligence also became dependent on the development of technical systems created exclusively for intelligence purposes, mainly in the field of signals intelligence and cryptanalysis, together with the employment of and exploitation of scientific expertise. This interaction between technology and scientific expertise from disciplines such as mathematics, statistics, and linguistics was a key element behind the British cryptologic successes against the German *Enigma* and a number of subsequent systems.[18] The Swedish success against the German *Geheimschreiber,* although facilitated by favorable circumstances and achieved with far more limited resources, was nevertheless accomplished through a remarkably smooth and nonbureaucratic employment of key scientific expertise for a defined intelligence purpose.[19]

Signal intelligence, and especially communications intelligence, or COMINT, stands out as a specific field where scientific expertise and

[16] The process of military bureaucratization and the friction it created is a lead theme in the literature on the Manhattan project. See Robert Jungk, *Brighter Than a Thousand Suns: A Personal History of the Atomic Scientists* (Harmondsworth: Penguin Books, 1982); Richard Rhodes, *The Making of the Atomic Bomb* (New York: Simon and Schuster, 1986); Silvan S. Schweber, "In the Shadow of the Bomb: Oppenheimer, Bethe, and the Moral Responsibility of the Scientist" (Princeton, NJ: Princeton University Press, 2000). On the parallel German case, see David C. Cassidy, *Uncertainty: The Life and Science of Werner Heisenberg* (New York: W. H. Freeman, 1992); Paul Lawrence Rose, *Heisenberg and the Nazi Atomic Bomb Project: A Study in German Culture* (Berkeley: University of California Press, 1998).

[17] The wide employment of radar led to new scientific demands. The performance of radar stations was calculated according to normal conditions, while the actual performance could vary according to different atmospheric conditions. In 1943, British and American scientists set up a joint committee to pool their knowledge of the phenomenon called propagation and to conduct further theoretical and experimental research to provide the armed forces with actionable propagation forecasts. See C. G Suits, George R. Harrison, and Louis Jordan, *Science in World War II. Applied Physics, Electronics, Optics, Metallurgy* (Boston: Little, Brown, 1948).

[18] Francis H. Hinsley and Alan Stripp, *Codebreakers: The Inside Story of Bletchley Park* (Oxford: Oxford University Press, 2001).

[19] On the breaking of the *Geheimschreiber* machine-crypto, see C. G. McKay and Bengt Beckman, *Swedish Signal Intelligence: 1900–1945* (London: Frank Cass, 2003).

methods were employed to solve an intelligence task. The prime mover was, as with all wartime research efforts, necessity; code breaking could not be accomplished on a wider scale or against high-grade systems without this influx from mainly the academic world, an influx that also included the non-conformity described at Bletchley Park—incidentally, the same phenomenon described in numerous accounts from the Manhattan project. Given this exclusive expertise, signals intelligence could not easily be subjected to the conformism of military bureaucracy and could retain an element of autonomy, as independent organizations or as a semi-autonomous profession.

Yet the mobilization of intellectual and institutional resources for military research and development also in itself constituted an emerging intelligence requirement. Academics were needed to monitor the work of colleagues on the other side. Shortly before the war, the young Briton with a doctorate in physics from Oxford, R. V. Jones, was approached by a staff member of Sir Henry Tizard's Committee for the Scientific Survey of Air Defence. Britain was, in this period, ahead of most other nations in integrating the academic scientific community and defense research and development, as illustrated by the lead in radar technology accomplished by the Royal Air Force (RAF) over the main adversary, Germany. However, the Committee had, as war drew closer, experienced a problem regarding intelligence, or rather lack of intelligence: the British intelligence services simply did not provide material that gave any insights into German efforts to apply science in aerial warfare.[20] Jones was offered the task of closing this gap and came to head one of the first forays in scientific intelligence. In the end, he found that task less challenging than scientific research: after all, the only thing he had to do was to figure out things that others already had discovered and thus by definition were achievable.[21] To some extent the growing use of science in warfare and in intelligence was, as Jones discovered, simply two sides of the same coin: scientific development of the means of warfare created the need for countermeasures, impossible without intelligence coverage of developments on the other side of the hill, an interaction that would continue and suffuse much of Cold War intelligence.

Scientific intelligence and technological intelligence were only two of a wide range of tasks that were rapidly expanding and transforming intelligence organizations. Foreign intelligence, cryptography, special operations, deception, and psychological warfare were tasks that not only demanded new agencies but also human resources with new competences. However, recruitment from the universities was more than adding specific competences to fields or problems

---

[20] R. V. Jones, *Most Secret War: British Scientific Intelligence 1939–45* (London: Coronet Books, 1979), p. 1.

[21] Ibid., p. 662.

that could not be handled without them. The new recruits joined intelligence not only as specialists but also as intelligence officers, serving along with officers with a more traditional background in the military. One very obvious reason for this kind of recruitment was the need for language skills in achieving a range of intelligence tasks, from espionage, or human intelligence (HUMINT) collection, to the study of open sources and the conduct of intelligence liaison and covert operations. The wartime U.S. Office of Strategic Services (OSS), as well as its British counterpart, the Secret Intelligence Service (SIS), or MI-6, needed people who spoke German but who also understood German society, culture, and mindset. The same requirement arose concerning Japan, Italy, and the Axis-allied countries as well as countries under occupation.

This was not only a concern for the major powers and their intelligence efforts. In Sweden, a secret intelligence service for HUMINT collection and special operations was established in the autumn 1939. It was headed by a military officer, but the majority of the small staff came from the universities. One of the recruits was Thede Palm, a doctor of theology and fluent in German, who was assigned the task of interrogating travelers arriving on the regular ferry lines from Germany, screening them for any observation with intelligence significance. Dr. Palm, as he was referred to, would take over as director after the war and run secret intelligence for another twenty years.

Languages, however, like physics and mathematics, were only additional supportive competences to the conduct of intelligence. But university graduates, whether in economy, classics, or theology, also brought along something else— a way to think, to deal with intellectual problems, and to structure information. While not as immediately useful as the ability to read and speak German, this more general intellectual competence had a significance that was soon obvious. In its field operations, the OSS experienced the value of academic training, and in his final report, Calvin B. Hoover, head of the OSS North Central European Division, covering Germany, the Soviet Union and Scandinavia, noted that from his experience, intelligence officers who lacked a university or college background did not perform well out on mission, since these operators often were left to solve or even formulate tasks on their own, without any detailed guidance from remote headquarters and over unreliable and slow communications. These field officers needed the perspective gained by a theoretical education to grasp the complexity of the conditions under which they had to operate and the kind of information they needed, not to mention assessing the crucial distinction between gossip and hearsay, on the one hand, and information that could be verified and documented on the other.[22]

---

[22] Calvin B. Hoover, "Final Report (No Date) Rg 226 (OSS), Entry 210 Box 436," (National Archives, College Park, MD). As an example of inadequate educational background, Hoover

Perhaps the most ambitious effort to merge academia and intelligence in the analysis phase of the intelligence process during the World War II was the OSS Research and Analysis branch (R&A), organized in 1941 and gradually dismantled after 1945, when OSS was reorganized and remnants of R&A were transferred to the State Department. Research and analysis both facilitated and benefited from the recruitment of key academic experts in disciplines and specific fields relevant to intelligence. This expertise, as well as its analytic products, often exceeded what other governmental agencies could provide, and the analytic process incorporated established academic editorial and quality-control procedures.[23] Moreover, R&A not only compiled available information and analyzed strategic and policy problems, but it also could provide input to the policymaking process on such key issues as policy toward postwar Germany and relations with the Soviet war-time ally.[24]

The failure of R&A to be fully accepted, and to achieve influence on a par with the intellectual quality of its output was mainly the result of bureaucratic rivalry and suspicions of politicization. Staffed to a large extent from Ivy League universities, the R&A branch met with suspicion as being "a hotbed of academic radicalism."[25] OSS, and especially R&A, tended to be regarded by the State Department as a competitor; as a result, the State Department, on the one hand, requested reports and the transfer of key staff members but on the other withheld or only reluctantly shared vital information—for instance, diplomatic cables from Moscow.

refers to a report received by one of the OSS agents in Switzerland who in great excitement reported that he had learned the secret by which the Germans produced gasoline from coal. But instead of the complicated industrial process he simply described the principle of adding so many atoms of hydrogen to so many atoms of carbon, well known to any chemical engineer. This particular report was actually disseminated and, as Hoover remarks, "aroused considerable amusement from some of the agencies which received it." The Hoover example could be straight out of Graham Greene's, *Our Man in Havana* (London: Heineman, 1958), all the more so because Greene's protagonist, James Wormwold, sold vacuum cleaners in Cuba. To earn additional money, he agreed to run spies for Britain and created an entirely fictitious network. At one point, he sent pictures of vacuum cleaner parts to London, calling them sketches of a secret military installation in the mountains.

[23] For this report by world-leading social scientists, see Raffaele Laudani, ed., *Secret Reports on Nazi Germany: The Frankfurt School Contribution to the War Efforts. Franz Neumann, Herbert Marcuse, Otto Kirchheimer* (Princeton, NJ: Princeton University Press, 2013).

[24] Petra Marquardt-Bigman, "The Research and Analysis Branch of the Office of Strategic Services in the Debate of US Policies toward Germany, 1943–46," *Intelligence and National Security* 12, no. 2 (1997): 91–100; Betty Abrahamsen Dessants, "Ambivalent Allies: OSS' USSR Division, the State Department, and the Bureaucracy of Intelligence Analysis, 1941–1945," *Intelligence and National Security* 11, no. 4 (1996): 722–753.

[25] Barry Kätz, *Foreign Intelligence: Research and Analysis in the Office of Strategic Services, 1942–1945* (Cambridge, MA: Harvard University Press, 1989).

Intelligence in Search of a Discipline?

The Cold War became a period of massive consolidation and development of the intelligence profession, but not along this initial methodological line. The years of the improvised merger between intelligence and academia were generally over, and the intelligence organizations established themselves as bureaucracies with a key role in national security, a component perceived as indispensable for the various levels in the Cold War security system—ranging from long-term planning and arms development, to disarmament negotiation, early warning, crisis management, and support for ongoing operations. But the focus was firmly on the collection-dissemination process established during World War II and constituting the intelligence foundation of deterrence. Intelligence was, as Allen Dulles stated in his book, *The Craft of Intelligence,* an activity for professionals knowing the nature of their trade and focused on the collection and dissemination of vital information. While dealing with collection in two chapters, Dulles hardly mentioned analysis and instead referred to the task of compiling vital information.[26]

The Cold War brought about a surge in intelligence requirements, not unlike the process during the Second World War. For the West, the inaccessibility of intelligence from behind the Iron Curtain, the emerging arms race, and the increasing complexity of related estimates forced intelligence organizations not only to develop the means for data collection but also to reach out for expertise on matters that lay outside existing core competences. In the beginning of the 1950s Swedish military intelligence received increased funding to reorganize and recruit in-house expertise in political science and economy. The secret intelligence service, Thede Palm's *T-kontorets,* hired a well-known academic economist to analyze the Soviet economy; this task later grew into a low-profile research institute, the Eastern Economic Bureau, a joint venture between secret intelligence and Swedish industry.[27] But this was merely outsourcing, not really incorporating academia or the scientific method into the concept of intelligence, which was still dominated by its military and police heritage. Only in the domain of signals intelligence did a dual process actually function, incorporating academic expertise into an intelligence structure, imbuing this structure with some elements of academic standards and scientific thinking.

The overall lack of academic or scientific methods reflects the *non-development* of intelligence analysis as applied multi- or transdisciplinary science. The Cold War was dominated by secret intelligence and the construction

---

[26] Allen Dulles, *The Craft of Intelligence* (New York: Harper & Row 1963), p. 154.

[27] Thede Palm, *Några Studier Till T-Kontorets Historia*, vol. 21, Kungl. Samgfundet for Utgivande Av Handskrifter Rorande, Stockholm, 1999.

of Babylonian towers of technical intelligence collection. Compared to the 1940s, the divide did not close, at least on an intellectual level, but instead widened as intelligence moved away from the scientific domain toward establishing itself as something fundamentally different, detached, and secluded.

Recruits with a solid academic background were still wanted and in some specific fields were badly needed. The best prospective analysts of intelligence on the Soviet economy were still trained researchers in economics, with the East Bloc as their main field of interest. But on the whole, intelligence did not appear as a very intellectually profitable arena for academics. Cold War intelligence was, in the East as well as in the West, a closed domain of bureaucratic, compartmented agencies, with limited cross-communication. Returning briefly to his old intelligence domain in the 1950s, R. V. Jones was disappointed by what he encountered. Scientific intelligence had, in his view, been engulfed by Whitehall bureaucracy, and in the process, the ethos of innovative problem solving had been lost.[28] On the other side of the Atlantic, the CIA Office of Scientific Intelligence was similarly shrouded in the equally devastating Washington intelligence turf battles.[29] Certainly, as the Cold War proceeded, scientific competence was increasingly needed in both West and East to comprehend and assess the technological development of the adversary and to develop the means for intelligence collection and data processing necessary to accomplish this task.

Scientific work in intelligence could of course be a rewarding career in itself or a steppingstone to a subsequent career in politics or public administration. But something fundamental was lacking, There is little testimony of intelligence work after Bletchley Park that praises it as *intellectually* rewarding in the academic sense, if for no other reason because of the limited or nonexistent possibility of publishing anything of significance for a subsequent academic career. In his classical work on strategic intelligence published in 1949, Sherman Kent, himself an academic recruit to the OSS Research and Analysis Branch, acknowledged the importance not only of the development of the sciences as a competence pool but also of the employment of scientific methods in a hypothesis-driven intelligence process.[30] In a 1955 article—published

---

[28] Jones, *Most Secret War: British Scientific Intelligence 1939–45*, and R. V. Jones, *Reflections on Intelligence* (London: Mandarin 1990). Intelligence veteran Michael Herman strongly disagreed with Jones, arguing that Jones failed to take into account the realities of the Cold War intelligence machinery, based as it was on efficient production lines, especially in the SIGINT domain: Herman, *Intelligence Power in Peace and War* (Cambridge: Cambridge University Press, 1996). Also see R. V. Jones, *Instruments and Experiences: Papers on Measurement and Instrument Design* (Hoboken, NJ: Wiley, 1988).

[29] See Jeffrey T. Richelson, *The Wizards of Langley. Inside the CIA's Directorate of Science and Technology* (Boulder, CO: Westview, 2002).

[30] Sherman Kent, *Strategic Intelligence for American World Policy* (Princeton, NJ: Princeton University Press, 1949).

typically enough in the classified CIA-journal *Studies in Intelligence*—Kent highlighted the need for what he called an intelligence literature. Such a litera-ture would deal with issues such as the purpose of intelligence, provide defini-tions of the terminology, and constitute an "elevated" debate. Without this, Kent argued, intelligence would not be able to establish a permanent institu-tional memory and never reach full maturity in comparison with academic disciplines.[31]

For decades, nothing very much came of this classified call to arms. With intelligence agencies maintaining a low or nonexistent public profile, unwilling to declassify or publish anything that might remotely jeopardize the protec-tion of sources, methods, and institutional information, the academy was also held at arm's length, with few researchers and university institutions devoted to the study of intelligence. To boot, when stray academics did venture out into this terrain, their efforts were, as in the Norwegian case, not always appreci-ated or rewarded.[32] There is little testimony that people who left the academy for intelligence experienced an increase in intellectual freedom. True, there were important though less visible links between various research institu-tions and intelligence, not least in the domain of technical R&D associated with means for technical collection and development of critical intelligence infrastructure. Also, fields of basic science with intelligence implications were funded and exploited, most of all in the United States.

It could be argued that by using the output of the research community as well as recruiting specialists from its ranks, Cold War intelligence had man-aged if not to merge with science, then at least to link up with and exploit it, with science serving mainly as an intellectual provider and intelligence as a consumer in terms of personnel and products. Yet the impact of this selec-tive integration had, contrary to all figures of recruitment and employment of academic knowledge and methods, surprisingly little effect on the intellectual nature of intelligence. More bluntly, the profession of intelligence analysis has not generally become a science based-profession as a result of this process of

---

[31] Sherman Kent, "The Need for an Intelligence Literature," *Studies in Intelligence* 1, no. 1 (1955), 1–8, available at https://www.cia.gov/library/center-for-the-study-of-intelligence/csi-publications/books-and-monographs/sherman-kent-and-the-board-of-national-estimates-collected-essays/2need.html.

[32] The Swedish historian Stig Ekman was in 1974 assigned by a parliamentary commission to write a report on the performance of the Swedish military intelligence during five crises during the 1950s and '60s. Ekman, who had been one of the senior historians in charge of a large-scale research project on Sweden during the Second World War, wrote an extensive report only to have it classified Top Secret; he was unable to retrieve his own manuscript for more than 20 years, and then only in a heavily sanitized version, which he eventually published: Stig Ekman, *Den Militära Underrättelsetjänsten. Fem Kriser under Det Kalla Kriget (the Military Intelligence. Five Crises dur-ing the Cold War)* (Stockholm: Carlsson, 2000).

selective integration. Rather, it has become a semi-academic profession with an increasing segment of frustrated and dissatisfied staff members, which is something quite different.[33] The profession of intelligence has cultivated the mystique of intelligence analysis for a variety of reasons. The "craft" of intelligence was not one described as taught but as experienced. Intelligence could only be learned through intelligence: outsiders, however gifted, remained outsiders, non-members of the secret guild.[34]

If intelligence analysis, as an intellectual activity, is not primarily science-based, what is it based on? Assessments don't just come out of nowhere, even if one study by a talented anthropologist quotes analysts claiming just that, along with hints of the existence of a mysterious hidden science. In medicine, the corresponding answer would be "established practice."[35] A clinical treatment could either be based on scientific results, according to a common standard based on validation, or be based on experience, the way things have always been done and where the effect has been repeatedly monitored, justifying the conclusion that the treatment works. "Established practice" is a model that has a long background in the military profession, in policing, in education, and in farming, just to take a few fields of experience-based human activity. But as medicine illustrates, experience-based established practice is not enough; indeed, it can sometimes be disastrous, especially in fields with rapidly developing technologies, expanding knowledge, or emerging new or transformed health risks.

Why did half a century of debate over the importance of a scientific dimension in intelligence analysis lead to such remarkably meager results? Why has not a field so rapidly developing and of such high priority as intelligence not transformed long ago in this direction as a continuation of professionalizing? What we thus should look for is perhaps not the incentives for a science of intelligence to develop but rather the reasons it failed to do so.

[33] This semi-academic character of the analytic profession in the US intelligence community is well reflected in the ethnographer Rob Johnston's report: Rob Johnston, "Analytic Culture in the US Intelligence Community: An Ethnographic Study" (Washington, DC: Center for Study of Intelligence,Central Intelligence Agency, 2005). A number of non-US examples of semi- or non-academic intelligence cultures are given in Phillip H. J. Davies, and Kristian C. Gustafson, *Intelligence Elsewhere: Spies and Espionage Outside the Anglosphere* (Washington, DC: Georgetown University Press, 2013).

[34] For further comments on perceptions of intelligence as profession, see Wilhelm Agrell, "When Everything Is Intelligence, Nothing Is Intelligence," in *Kent Center Occasional Papers* (Washington, DC: Central Intelligence Agency, 2003).

[35] For a discussion of the scientific nature of medicine and the implications for intelligence, see Walter Laqueur, *A World of Secrets: The Uses and Limits of Intelligence* (London: Weidenfeld and Nicolson, 1985), p. 302. Also see Stephen Marrin and Jonathan D. Clemente, "Modeling an Intelligence Analysis Profession on Medicine 1," *International Journal of Intelligence and CounterIntelligence* 19, no. 4 (2006): 642–665.

The first and most important of the missing incentives is perhaps the self-image of the intelligence profession. The craft or mystery conception is not only a product of the absence of alternatives, of possible paths toward scientifically based analytic methods, but also a powerful instrument that intelligence analysts and officials use to draw a sharp dividing line between insiders and outsiders, those in the know and those not in the know and thus by definition unable to add something of substance. The impact of this for self-esteem, professional identity, and policymaker access should not be underestimated. The notion of a secret tradecraft is a powerful instrument for averting external critics, a method that can be observed not only in intelligence but also in many academic controversies, where interference by representatives from other disciplines often is far from well received, and almost by definition those who interfere are regarded as ignorant and irrelevant. The transformation of intelligence analysis toward overt, systematically employed, and verifiable methods would not completely remove but would inevitably weaken the protective wall surrounding craft and mystery. One of the most important aspects of the critical public debate since 2001 over the performance of intelligence has been the penetration of this wall, and thus possibly the weakening of this major negative incentive.

However, the unprecedented openness about analytic products and processes around the Iraqi WMD case has also affected a second negative incentive: the impact of secrecy. Secrecy, as a phenomenon in intelligence, is both functional and dysfunctional—both an obvious necessity to protect sensitive sources, methods, and valuable intellectual property and an element in the intelligence mythology employed to shield organizations and activities and to amplify the assumed significance of trivial information and flimsy assessments. The extensive employment of secrecy has, intentionally or not, blocked the intellectual development of intelligence analysis by drastically limiting the empirical basis for any such process. True, there is a rapidly expanding scholarly literature on the history of intelligence, based on a growing range of documentary sources, but there is a vast time lag between the periods covered by the historians and the contemporary conduct of intelligence analysis. Furthermore, the documentary material available to historians is often incomplete and in some cases misleading due to prevailing secrecy.[36]

Intelligence analysis cannot abandon secrecy for the sake of methodology. But secrecy can be employed in a more selective and "intelligent" way if the empirical studies and methodological self-reflection are regarded not as an external concern but as being in the interest of intelligence itself. Rob

---

[36] See Richard J. Aldrich, "Policing the Past: Official History, Secrecy and British Intelligence since 1945," *English Historical Review* 119, no. 483 (2004).

Johnston's study is an interesting example of a development in this direction, applying the tools and perspectives of an anthropologist to the processes of intelligence analysis.[37] Still, a similar exercise would probably be regarded with horror by most intelligence agencies around the world.

Yet the negative incentives blocking the employment of scientific methods and perspectives in intelligence analysis cannot be ascribed only to the intelligence culture and the modus operandi of intelligence agencies. The wider political, public, and not least academic context has also reinforced the intellectual isolation of intelligence. There has been little or no political interest in stimulating contacts over the university-intelligence divide. Political self-preservation has for long periods been a strong incentive against any such enterprise; if penetration of intelligence agencies by hostile services was a menace, academic penetration was only a variant and in some respects worse, since the net result could be not only the penetration of secrets but the open disclosure of them. And as for the academic communities themselves, there has been a widespread reticence toward intelligence organizations, sometimes developing into hostility—and not always without reason. Somewhat more trivial, but not less important, is a widespread lack of interest in intelligence matters from the academic disciplines concerned, due partly to academic priorities, partly to ignorance or (more or less well-grounded) mistrust. Academic interest in intelligence matters will come about only when intelligence is perceived as an interesting and rewarding field for research, a process now slowly under way, though by no means on any grand scale. The rise of intelligence studies will not automatically transform intelligence analysis, but it can reduce the academic ignorance and possibly some of the hang-ups in the interrelationship between the two cultures.

Science in Search of a Problem-Solving Role

Driven by external impulses not unlike those shaping the merger between the scientific world and the military in the 20th century, scientific research has become increasingly engaged in handling major problems regarding risk and security in a wider social context. If war could force science down from the ivory tower, then risk in post-modern societies seems to continue that process. This has been especially evident in environmental issues but has also permeated most aspects of what the German sociologist Ulrich Beck has termed "risk society," in which social and technological developments create increasingly complex

---

[37] Johnston, "Analytic Culture in the US Intelligence Community."

and unmanageable mega-risks, with citizens demanding an increasingly unattainable security.[38] Increasing, or increasingly visualized, societal risks have resulted in a demand-pull from policy, commercial interests, and opinion and pressure groups. Yet there is also an element of incentive push from the scientists themselves, a complex interaction of personal and professional ambitions, temptations to tread ground not already covered and the ethos of making a contribution to public good. Intelligence analysts, though in many instances similarly motivated by an over-arching mission,[39] are considerably less likely to be able to successfully exercise any such incentive push, given the institutional context of far less room for maneuvering and fewer initiatives than in academia.

The transformation of science from the autonomous endeavors dominating universities in the 19th century to research increasingly conducted on behalf of social actors and financed by them was gradual, and as described earlier, was greatly influenced and in many respects shaped by the two world wars and the subsequent Cold War arms race. Science, once mobilized for over-arching societal goals, was never demobilized, though the purpose of mobilization gradually changed. Vannevar Bush's famous report, *Science the Endless Frontier,* written against the background of the wartime experiences of large-scale goal-oriented research, came to symbolize this wider call to arms, where science, in a new postwar world, should take on such peacetime tasks as battling diseases and improving public welfare.[40] From the 1950s onward these two streams of research, pure science and goal-oriented research, coexisted in various mixes, accompanied by an ongoing debate over the balance between scientific autonomy and usefulness in a social context.[41] An increased orientation toward the society and problems formulated in the political arena or in the market did not immediately alter the disciplinary structure or the way research was carried out, but gradually the difference began to be more visible.

In *The New Production of Knowledge,* Michael Gibbons defined two types or modes of research.[42] Traditional research, where knowledge is generated within a disciplinary, primarily cognitive, context was designated Mode 1,

[38] Ulrich Beck, *Risk Society: Towards a New Modernity* (London: Sage, 1992). For a wider discussion on science and risks, see Maurie J. Cohen, *Risk in the Modern Age: Social Theory, Science and Environmental Decision-Making* (Basingstoke: Palgrave, 2000).

[39] Michael Herman identifies four characteristics of people belonging to an intelligence culture, a sense of being different, of having a distinct mission, the multiplying effect of secrecy, and finally that of mystery. Herman, *Intelligence Power in Peace and War,* pp. 327–329.

[40] Vannevar Bush, "Science: The Endless Frontier," *Transactions of the Kansas Academy of Science (1903–),* 48, no. 3 (1945).

[41] For the development of postwar research policy in the OECD countries, see Salomon, *Science and Politics.*

[42] Michael Gibbons et al., *The New Production of Knowledge: The Dynamics of Science and Research in Contemporary Societies* (London: Sage, 1994).

while a new form of research is carried out in the context of applications. While Mode 1 is characterized by epistemological homogeneity, Mode 2 is heterogeneous, trans-disciplinary, and employing a different type of quality control. In a sense, Gibbons's often-quoted model simply reflected the innovation in research structure in American, British, and German wartime research. The difference was that Mode 2 now had expanded beyond the realm of national security and reflected the wider role of science heralded by the utopian endless frontier and the dystopian risk society. In Mode 2 problems are defined in a different way, one more inclusive to the practitioners and demanding a multidisciplinary approach.[43] Fields such as environment, climate, or migration are simply too broad and too complex for a single academic discipline. The emerging process toward multi-disciplinary research, starting internally in the academy in the 1960s, thus met an external demand-pull from the 1990s and onward. The implication of Mode 2 is a transformation of research in terms of problem definitions, ways to operate, and not least links with external interests, the stakeholders—a transformation that in many respects tends to make research less different from intelligence. This convergence can be observed especially where scientific expertise and research efforts are directly utilized in addressing major risks in the society or are mobilized in crisis management.[44] We pursue these commonalities between intelligence and what might be called "policy analysis 2.0" in more detail in Chapter 6.

A New Discipline—Or Common Challenges?

In 1971, Harvard professor Graham T. Allison published what was to become one of the most famous and widely used methodological textbooks on crisis management and foreign policy decision making, *Essence of Decision. Explaining the Cuban Missile Crisis.*[45] The book was not a historical account of the climax of the Cold War but rather a methodological postmortem of a crisis, employing the surgical instruments of social sciences. In three "cuts" Allison applied three models for analyzing foreign policy decision making. Model I, the rational actor model, assumes that states act strategically, trying to maximize gains and minimize risks; an international crisis thus could take the form of interacting games. Model II, the organizational process, also

---

[43] Ibid., pp. 3–8.

[44] One of very few discussions of the intelligence aspects of risk science working on the basis of a preventive paradigm is Mark Phythian, "Policing Uncertainty: Intelligence, Security and Risk," *Intelligence and National Security* 27, no. 2 (2012): 187–205.

[45] Graham T Allison, *Essence of Decision: Explaining the Cuban Missile Crisis* (Boston: Little, Brown, 1971).

assumes rationality, though on a different level. Here actions are determined not by choices between strategies but by the operations and repertoires of state bureaucracies. Decisions are not taken; they are produced by organizational momentum, and the search for model I rationality would be futile or misleading. Model III also focuses on the sub-actors, in this case not structures but individuals and groups; governmental politics highlight the kind of corridor politics where the outcome would be determined by power games and shifting coalitions far from abstract national interests or bureaucratic logic. In a second edition, published in 1999, new historical findings were incorporated, but those changed remarkably little of the analysis. More important was perhaps the addition of a new aspect of model III: Irving L. Janis's theory of groupthink, where the outcome would not be determined by the interaction and goals of sub-actors but by a specific form of self-regulating psychological group behavior.[46]

While *Essence of Decision* was not about intelligence analysis per se, the theme had a number of obvious intelligence dimensions. First, Allison had transformed a real-time intelligence problem into a retrospective scholarly problem; some of the issues investigated were the same questions that US and Soviet intelligence analysts had struggled with before and during the crisis, such as the assumed rationality of the opponent and possible responses down the road toward confrontation. The major difference was the perspective and employment of methods; Allison convincingly showed that the choice of analytic models to a large extent determined the result. There was not *one* explanation but several parallel ones, making the account a splendid seminar piece. Depending on your choice of method, you would end up with different explanations and explain different aspects of the crisis, while some questions would remain unsolved and possibly unsolvable—the secrets and mysteries in intelligence analysis.

To be sure, the seminar piece would not have been very helpful for intelligence consumers during the crisis. The employment of alternative and seemingly convincing analytic approaches would not satisfy any intelligence requirements. Questions like "Will they launch the missile if we strike them?" cannot be handled through a methodological exercise and "on the one hand, but on the other hand" arguments. Still, the logic of the seminar piece is often there, though not explicit and visible. While intelligence is often conducted in an a-theoretical and fact-based way—in accordance with the assumption that intelligence, in contrast to academic scholarship, is about "reality"—the theories are always reflected

[46] Graham T. Allison and P. Zelikow, *Essence of Decision: Explaining the Cuban Missile Crisis* (New York: Longman, 1999); Irving L. Janis, *Groupthink: Psychological Studies of Policy Decisions and Fiascoes* (Boston: Houghton Mifflin, 1972).

in one way or another. The famous US Special National Intelligence Estimate drafted by the CIA on September 19, 1962,[47] just before the crisis erupted, is not an example of mystifying intelligence tradecraft but an almost textbook like application of the rational actor model on the issue of a possible Soviet choice to deploy nuclear ballistic missiles on Cuba—an alternative discarded with an array of logical arguments by the top CIA analysts.[48]

In much the same way, attempts by Western intelligence agencies to assess the intentions and strategic choices of Saddam Hussein from the summer 1990 and onward were based on the assumption that the Iraqi president was acting rationally, something finally cast into doubt by the lengthy and confused statements made by Hussein to his interrogators while in custody. Even more surprising, perhaps, was the failed attempt to reconstruct the organizational process behind the assumed or aborted Iraqi WMD program. It was not only the weapons and facilities that were missing; the archival excavations of the Iraqi Survey Group failed to uncover such an organizational process. There was no paper trail of important matters; instead, the major governmental agencies seemed to have been preoccupied with peripheral issues and the collection of personal information on their employees.[49] Iraq was not a failed state; rather, what the excavation uncovered was a *fake* state, with the dysfunctional governmental apparatus typical for a totalitarian state guided by fear. Problems in providing intelligence assessments cannot be solved by just bringing in the researchers or outsourcing the analytic tasks. Depending on the disciplinary affiliation and the school they belong to, the researchers will tend to come up not only with different questions but also different answers to the same question. And the intelligence relevance of these answers will inevitably vary.

Scientific researchers and intelligence analysts both try to comprehend the world, sometimes in a strikingly similar way, but sometimes with different methods and using different material. A scientific analysis can appear cumbersome and esoteric from an intelligence perspective, and correspondingly, intelligence can appear as sloppy research, the gathering and analyzing of incomplete data. The single U-2 over-flight of Cuba on October 14, 1962,

[47] The Military Buildup in Cuba, SNIE 85-3-62, 19 September 1962, reproduced in part in Mary S. McAuliffe, *CIA Documents on the Cuban Missile Crisis, 1962* (Washington, DC: Central Intelligence Agency, 1992).

[48] The final twist to this misconception was supplied by Sherman Kent, the most senior analyst responsible for the drafting of the SNIE, who after the event argued that it was not the CIA analysts who had been wrong but Nikita Khrushchev, since the former had foreseen disaster for the Russians if they tried such a threatening move against the United States. See Raymond L. Garthoff, "U.S. Intelligence in the Cuban Missile Crisis," in James G. Blight and David A. Welch, eds. *Intelligence and the Cuban Missile Crisis* (London: Frank Cass, 1998).

[49] Charles A. Duelfer, Comprehensive Report of the Special Advisor to the DCI on Iraq's WMD. Washington, September 30, 2004, available at https://www.cia.gov/library/reports/general-reports-1/iraq_wmd_2004/.

certainly failed to meet elementary scientific demands for validity, statistical significance, and blind tests. No drug would have been prescribed, let alone certified, on such a flimsy basis. But the single over-flight, and the disclosure of excavations and trucks for fluid oxygen on the aerial photographs, was enough to verify the *suspicion* about Soviet missiles. In intelligence "How much is enough?" is a question about time, resources, and risk. In research, it is a question of scientific reliability and the display of methodological rigor.

Some of the problems in intelligence have little or no relevance from an academic perspective. And correspondingly, many research problems lack an intelligence dimension or significance. But then there are the fields that overlap, where intelligence and research try to comprehend a common phenomenon, though with a difference in purpose. One such case was the rapidly accelerating crisis in the global financial system in 2008. The crisis was not a bolt out of the blue—rather the contrary. Theories of international economy foresee the inevitability of recurring recessions. Still, this one came as a surprise, much in the way Ephraim Kam describes the phenomenon of surprise attacks; what happens is often not something completely inconceivable but a development that in principle had been foreseen but that appeared at the wrong time, the wrong place, in the wrong way.[50] The crisis in the global financial system challenged existing economic theories not so much due to the lack of actionable early warning but rather due to the difficulties in comprehending and explaining the dynamics, spread, and impact of the crisis. The crisis was a crucial intelligence problem—for all economic actors, for governments, financial institutions, and companies. But the crisis also constituted a major intelligence challenge for the actors outside the economic sphere, the foreign and domestic intelligence services tasked with monitoring the possible social and political impact of the crisis. A recession of this magnitude would inevitably change the world, and those monitoring it had to realize that the script was being rewritten. But how should this process be comprehended? And what models could be used?

While a world economic crisis constitutes a huge common intellectual problem for researchers and intelligence analysts alike, other fields constitute very different kinds of common problems. Cultural understanding in the context of counterterrorism and counter-insurgency is a recurring and in retrospect often embarrassing intelligence failure. But in contrast to the world economic crisis, it is not in the same way a shared intellectual problem between intelligence and research. Intelligence is, for reasons of resources and demands, not concerned with cultural studies as such; cultural studies for intelligence purposes are

---

[50] Ephraim Kam, *Surprise Attack: The Victim's Perspective* (Cambridge, MA: Harvard University Press, 1988).

much more like that single U-2 over-flight, guided by specific intelligence relevance but also by the level of reliability needed to supply actionable answers. From a research perspective the issues at stake for intelligence might appear as uninteresting or irrelevant, and the methods employed as dubious. There is, however a complication not only in the methods and the way studies are designed. The major controversy is one of purpose, the use or possibly misuse of research.[51] If research on intelligence might be provoking, conversely, so is intelligence as research. To what extent do researchers want their methods and findings to be employed for intelligence purposes? And how far can academic research be pursued before colliding with secrecy and the interests of national security? Here the issue at stake is not primarily one of differences in interpretations, but one of research and intelligence ethics.

Intelligence analysis and research can thus, as those drafting or studying intelligence assessments over and over again have observed, share methods and be confronted with similar or even identical problems. But are they two sides of the same coin? Or are they the same side of two different coins? To be more precise, is the divide about two different social contexts or two basically different principles to construct knowledge, though sometimes with similar methods? This highlights a fundamental underlying issue: whether intelligence analysis and scientific research can be defined as variants of a single epistemology, or whether they represent two separate epistemologies, not only in terms of organization and culture but also in terms of how knowledge is perceived and created. Looking back on the divide that in many respects was a creation of the Cold War security and intelligence universe, the answer would be the latter. However, looking at more recent experiences, and the outlook for the future, the answer would be a different one. What we are observing is something that within the sociology of science is called epistemological drift where the two coins are merging, not completely, but in this fundamental respect.

---

[51] For a classical study on the issue, see Irving Louis Horowitz, ed., *The Rise and Fall of Project Camelot: Studies in the Relationship between Social Science and Practical Politics* (Cambridge, MA: MIT Press, 1967).

3

What Is Analysis? Roads Not Taken

Given Cold War history and the sheer scale of the US intelligence enterprise, the United States has cast a long shadow over the practices of its friends and allies. This chapter explores what intelligence analysis is, and why, especially in American practice, it has turned out as it has. Yet analysis is hardly singular; there are many forms for many customers with many different needs. Moreover, trying to understand analysis on its own is listening to one hand clapping: it cannot be understood apart from what are still—and unhelpfully—called consumers (or worse, customers). Nor can analysis be separated from collection. More than a half century after publication, the writings of Sherman Kent and his critics are vivid about the basics of US analysis and equally vivid about what might have been. Intellectually, if perhaps not politically, intelligence analysis did not have to distinguish sharply between "foreign" and "domestic," with the trail of foreign assessment stopping at the water's edge. Nor did it have to give pride of place to collection, with the first question asked about uncertain judgments being: can we collect more? Nor did it have to separate intelligence more sharply than did most countries from policy and politics lest intelligence become "politicized."

What Is Analysis?

Start with the basics, the nature of the question or issue, and what are called— not very helpfully as will be seen later—consumers. Table 3.1 lays out three categories of questions, from puzzles, to mysteries, to complexities.[1] When the Soviet Union would collapse was a mystery, not a puzzle. No one could know

---

[1] On the distinction between puzzles and mysteries, see Gregory F. Treverton, "Estimating beyond the Cold War," *Defense Intelligence Journal* 3, no. 2 (1994), and Joseph S. Nye Jr, "Peering into the Future," *Foreign Affairs* (July/August 1994): 82–93. For a popular version, see Gregory F. Treverton, "Risks and Riddles: The Soviet Union Was a Puzzle. Al Qaeda Is a Mystery. Why We Need to Know the Difference," *Smithsonian* (June 2007).

Table 3.1 **Intelligence Puzzles, Mysteries, and Complexities**

Type of Issue	Description	Intelligence Product
Puzzle	Answer exists but may not be known	*The* solution
Mystery	Answer contingent, cannot be known, but key variables can, along with sense for how they combine	Best forecast, perhaps with scenarios or excursions
Complexity	Many actors responding to changing circumstances, not repeating any established pattern	"Sensemaking"? Perhaps done orally, intense interaction of intelligence and policy

the answer. It depended. It was contingent. Puzzles are a very different kind of intelligence problem. They have an answer, but we may not know it. Many of the intelligence successes of the Cold War were puzzle-solving about a very secretive foe: were there Soviet missiles in Cuba? How many warheads did the Soviet SS-18 missile carry?

Puzzles are not necessarily easier than mysteries: consider the decade required to finally solve the puzzle of Osama bin Laden's whereabouts. But they do come with different expectations attached. Intelligence puzzles are not like jigsaw puzzles in that we almost certainly won't have all the pieces and so will be unsure we have the right answer. The US raid on Osama bin Laden in 2011 was launched, according to participants in the decision, with odds no better than six in ten that bin Laden actually was in the compound. But the fact that there is in principle *an answer* provides some concreteness to what is expected of intelligence. By contrast, mysteries are those questions for which there is no certain answer. They are iffy and contingent; the answer *depends* not least on the intervention, be it policy or medical practice. Often, the experts—whether intelligence analysts, doctors, or policy analysts—find themselves in the position of trying to frame and convey essentially subjective judgments based on their expertise.

"Complexities," by contrast, are mysteries-plus.[2] Large numbers of relatively small actors respond to a shifting set of situational factors. Thus, they do not

---

[2] The term is from David Snowden, "Complex Acts of Knowing: Paradox and Descriptive Self-Awareness," *Journal of Knowledge Management* (2002) 6, no 2. His "known problems" are like puzzles and his "knowable problems" akin to mysteries.

necessarily repeat in any established pattern and are not amenable to predictive analysis in the same way as mysteries. Those characteristics describe many transnational targets, like terrorists—small groups forming and reforming, seeking to find vulnerabilities, thus adapting constantly, and interacting in ways that may be new. Complexities are sometimes called "wicked problems," and one definition of those problems suggests the challenges for intelligence, and in particular the "connectedness" of the threat with our own actions and vulnerabilities:

"Wicked problems are ill-defined, ambiguous and associated with strong moral, political and professional issues. Since they are strongly stakeholder dependent, there is often little consensus about what the problem is, let alone how to resolve it. Furthermore, wicked problems won't keep still: they are sets of complex, interacting issues evolving in a dynamic social context. Often, new forms of wicked problems emerge as a result of trying to understand and solve one of them."[3]

The second thing to notice about intelligence analysis is that it is plural. The "analysis" done in translating a video image from a drone into target coordinates that appear in a pilot's cockpit—what the Pentagon calls DIMPIs (designated mean points of impact, pronounced "dimpy")—may be totally a processing operation and all done automatically, without human hands (or brains) in the process once the system is designed. At the other end of the spectrum, deep mysteries, like charting Egypt's future after the "Arab spring" in 2011, require several kinds of human expertise and will be enriched by employing a variety of analytic techniques. In between, there are in principle a multitude of needs that consumers have for analysis. Table 3.2 identifies a dozen kinds of needs to make that multitude manageable.

For each need, the table identifies whether the question at issue is a puzzle, a mystery, or a complexity. It also hazards guesses for each, about how much demand there will be from policy officials and how much time will be required of those officials in framing and conveying the intelligence. For instance, self-validating tactical information, those DIMPIs again, are a puzzle in high demand from policy officials or operators and don't require much time from those officials. They are self-validating in the sense that the target either is or isn't where the coordinates say it should be, and exactly how the coordinates were put together is of scant interest to the consumer.

By contrast, if the task asked of analysis was to assess the implications of various policy choices, that process would be mystery-framing, not puzzle-solving. It would take both time and candor by policy officials who would have to tell analysts what alternatives were under consideration in a way that seldom

---

[3] Tom Ritchey, *Structuring Social Messes with Morphological Analysis* (Stockholm: Swedish Morphological Society, 2007), pp. 1–2.

Table 3.2 **Intelligence Needs, Types, and Demands**

Intelligence Need	Type of Issue	Likely Demand from Policy Officials	Demands on Officials' Time
Self-validating tactical information	Puzzle	High	Low
Warning with pre-developed indicators	Mystery turned into puzzle	High	Medium
Warning with more subjective indicators	Mystery	Medium	Medium
Tactical nonmilitary support	Puzzle, sometimes mystery	High	Low to medium
"Connecting the dots"	Puzzle, sometimes mystery	Medium to high	Medium
Categorizing emerging issues	Mystery	Medium, but small time window	Medium
Assessing the implications of policy choices	Mystery	High if agree, low otherwise	High
Framing the future	Mystery	Low	Medium
Expert views of deep mysteries or complexities	Mystery, complexity	Low	Medium
"Sensemaking" about the future	Complexity	Unknown	High

happens. And it is perhaps not too cynical to observe that policy officials are more likely to want this sort of analysis if they think it will support their preferred position. Trying to make sense of complexities probably requires even more policymaker time, for it should be done jointly, perhaps in a table-top exercise where all hypotheses and all questions are in order given the shapelessness of the issue.

Conveying Uncertainty

The complexities example drives home the point that uncertainty cannot be eliminated, only assessed and then perhaps managed. That is more and more obvious when the analytic task moves away from warning, especially very tactical warning, toward dealing with more strategic and forward-looking mysteries, one for which the analysis begins where the information ends and uncertainty is inescapable. In framing this task, it is useful to compare Carl von Clausewitz with his lesser-known contemporary strategist, Antoine-Henri, Baron de Jomini.[4] Jomini, a true child of the Enlightenment, saw strategy as a series of problems with definite solutions. He believed that mathematical logic could derive "fundamental principles" of strategy, which if followed should mean for the sovereign that "nothing very unexpected can befall him and cause his ruin."[5] By contrast, Clausewitz believed that unpredictable events were inevitable in war, and that combat involved some irreducible uncertainty (or "friction"). He characterized war as involving "an interplay of possibilities, probabilities, good luck and bad," and argued that "in the whole range of human activities, war most closely resembles a game of cards."[6]

Intelligence, perhaps especially in the United States, talks in Clausewitzian terms, arguing that uncertainty, hence risk, can only be managed, not eliminated. Yet the shadow of Jomini is a long one over both war and intelligence. In fact, intelligence is still non-Clauswitzian in implying that uncertainty can be reduced, perhaps eliminated. That theme runs back to Roberta Wohlstetter's classic book about Pearl Harbor, which paints a picture of "systemic malfunctions."[7] There were plenty of indications of an impending attack, but a combination of secrecy procedures and separated organizations kept them from being put together into a clear warning. If the dots had been connected, to use a recently much-overused phrase, the attack could have been predicted. So, too, the United States report on the terrorist attacks on 9/11 imposes a kind of Wohlstetter template, searching for signals that were

---

[4] This discussion owes much to Jeffrey A. Friedman and Richard Zeckhauser, "Assessing Uncertainty in Intelligence," *Intelligence and National Security* 27, no. 6 (2012): 824–847.

[5] Antoine Henri Baron de Jomini, *The Art of War*, trans. G. H. Mendell and W. P. Craighill (Mineola, NY: Dover, 2007), p.250.

[6] C. Von Clausewitz, *On War*, trans. Michael Howard and P. Paret (Princeton, NJ: Princeton University Press, 1976). For a nice comparison of Clausewitz and Jomini, see Mark T. Calhoun, "Clausewitz and Jomini: Contrasting Intellectual Frameworks in Military Theory," *Army History* 80 (2011): 22–37.

[7] Roberta Wohlstetter, *Pearl Harbor: Warning and Decision* (Palo Alto: Stanford University Press, 1962).

Table 3.3 **Jominian versus Clausewitzian Intelligence**

Jominian	Clausewitzian
Goal is to eliminate uncertainty	Goal is to assess uncertainty
There is a "right" answer	"Fog of war" is inescapable
More information and better concepts narrow uncertainty	Single-point high-probability predictions both unhelpful and inaccurate
Large uncertainty indicates shortcomings in analysis	Better analysis may identify more possible outcomes

present but not put together.[8] The perception of linearity is captured by its formulation "the system is blinking red." Table 3.3 summarizes the differences between the Jominian and Clausewitzian approaches:

The Jominian approach pervades how analysis is done and how it is taught. Most assessments, like American National Intelligence Estimates (NIEs), provide a "best" estimate or "key judgments." They may then set our alternatives or excursions, but the process tends to privilege probability over consequences, when in fact it is the combination of the two together that matters to policy. This emphasis on "best bets" also runs through familiar analytic techniques, like analysis of competing hypotheses (ACH). But "competition for what?" The usual answer is likelihood. Indeed, the original description of ACH, in the now-classic book by Richards Heuer, explains its goal as being to determine "which of several possible explanations is the correct one? Which of several possible outcomes is the most likely one?"[9]

A true Clausewitzian approach would rest, instead, on three principles:

- Confidence and probability are different, thus there is no reason not to be explicit about probabilities, even with low confidence.
- Content of information matters as much as reliability, so, again, important information should not be excluded simply because it is deemed not reliable.
- Perhaps, most important, consequence matters, in evaluating information and in constructing alternatives; thus consequential possibilities should not be relegated to the sidelines because they are judged unlikely.

[8] National Commission on Terrorist Attacks upon the United States, *The 9/11 Commission Report: Final Report of the National Commission on Terrorist Attacks upon the United States* (Washington, DC, 2004). Available at http://www.9-11commission.gov/.

[9] Richards J. Heuer, *Psychology of Intelligence Analysis* (Washington, DC: Center for the Study of Intelligence, Central Intelligence Agency, 1999), p. 95.

The resulting product would, in effect, lay out a probability distribution—not multiple answers but rather a single distribution. If consequence were deemed as important as likelihood, intelligence would, in effect, produce a probability distribution in which consequential outcomes would receive attention not just as excursions, even if their probability was low or could not be assessed very clearly. In looking at 379 declassified US NIEs, Friedman and Zeckhauser found but one example of this style of analysis. A 1990 NIE, *The Deepening Crisis in the USSR,* laid out on a single page four different "scenarios for the next year" in a simple figure.[10] Each was explained in several bullets, and then assessed a "Rough Probability." The scenario deemed most likely was presented first but not given any more discussion than the others. The NIE thus neglected neither probability nor consequence. It conveyed no sense that one scenario should be thought of as "best" or "correct." Nor did it require readers to parse meaning of concepts like "significant," "serious," or "important" (even if those were elaborated in a glossary, as is now the practice for NIEs). In the end, it allowed readers to decide for themselves which possibilities deserved pride of place.

Yet the question of whether busy senior policymakers would sit still for a Clausewitzian approach is a fair one. The easy answer would be to try it as an experiment for a handful of estimates on issues that are both important and very uncertain. A hint of an answer was provided by Stephen Hadley, President George W. Bush's national security advisor, based on his experience of the November 2007 NIE on Iran's Nuclear Intentions and Capabilities discussed in more detail later. Hadley made the intriguing—and Clausewitzian in spirit—suggestion that for the several most important, and uncertain, issues a president faces, intelligence might present its assessment in ways different from the "we judge with medium confidence" format. Imagine, for example, a pie chart representing all the United States would like to know about an issue. Different slices might be different sizes based on judgments about how important they are. In this case, the "is Iran weaponizing?" slice would have been significant but smaller than the "is it enriching?" piece, since the latter paces the country's nuclear program. And the slices might show clearly how much US intelligence knew about that piece. The weaponization slice would have indicated good information on that score.

In effect, the new information provided the solution to a puzzle—where does Iran's weaponization program stand, or at least where did it stand circa 2003? Intelligence puzzles have an answer, but, as in this case, the answer may be known to the target but not the interested state, like the United States. The state of Iran's enrichment program included both puzzles and mysteries. The technical aspects

---

[10] Friedman and Zeckhauser, "Assessing Uncertainty in Intelligence," p. 832.

could be thought of as puzzles: how many centrifuges with what capacity and so on? Yet the critical questions were mysteries: what did Iran intend with its enrichment program? What were the critical determinants of decisions about it? And critically, how would Iran respond to various sticks and carrots offered by the international community? With regard to weaponization, the NIE inferred a conclusion about that last mystery from the puzzle it had solved: Iran's leaders had stopped its weaponization program at least partly in response to international pressure.

Turning mysteries into puzzles is a temptation in analytic tradecraft. That was a conspicuous feature of the October 2002 NIE on Iraq. The question had narrowed to a puzzle: does Saddam have weapons of mass destruction? There was not much "Iraq" in the NIE; even the dissents turned on technical matters, like aluminum tubes. A Clausewitzian approach can hardly eradicate that temptation, but it might help lay out more clearly the puzzle and mystery portions of the issue being assessed, and serve as a check on neglecting important mysteries simply because we don't know much about them.

The Fallacy of the Intelligence Cycle: Primacy to Collection

Most critiques of the canonical intelligence cycle, depicted in Figure 3.1, emphasize its shortcomings as a description of how intelligence and its connection to policy actually work. For instance, for long-time CIA analyst Arthur Hulnick, "When it came time to start writing about intelligence, a practice I began in my later years at the CIA, I realized that there were serious problems with the intelligence cycle. It is really not a very good description of the ways in which the intelligence process works."[11] The critiques are apt, usually focusing on the pressure of time and thus the overlapping or skipping of steps. Policy officials have neither time nor much ability to be very clear about what they'd like from intelligence; a little like the reverse of US Supreme Court Justice Potter Stewart on pornography, they'll know the good stuff when they see it.[12] As a result, collection proceeds even as requirements are still being developed. So, too, the cycle skips steps all the time, thus the dotted lines in the figure: consumers receive "raw" intelligence or drafts or briefings on work in progress. "Dissemination" thus takes many

---

[11] Arthur Hulnick, "What's Wrong with the Intelligence Cycle," in *Strategic Intelligence* ed. Loch Johnson (Westport CT: Greenwood, 2007), p. 1.

[12] Stewart's line was "I know it when I see it." F. P. Miller, A. F. Vandome, and M. B. John, *Jacobellis v. Ohio* (Saarbrücken, Germany: VDM Publishing, 2011).

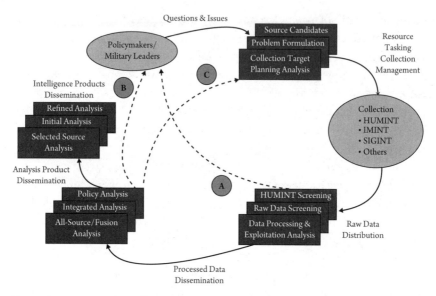

Figure 3.1 Canonical Intelligence Cycle.

forms in addition to the finished products emphasized by the intelligence cycle. Consumers ask questions, which in turn touch off new collection. So the critiques mostly imply that while the canonical intelligence cycle may be a kind of ideal type, it is too abstracted from the reality of real-life intelligence to be useful.

Yet, in fact, the cycle is worse than just a too-abstract ideal. It is wrong in principle, not just application. To the extent that its logic, if not its details, drive the work of intelligence, it is a positive impediment to doing better. Not only does it turn what is, or should be, a process of continual iteration between intelligence and policy into a sequential one but it also gives pride of place to intelligence *collection*. The privileging of collection in the cycle is very Jominian. It presumes that more information will reduce uncertainty and so let intelligence come nearer to the truth: recall Wohlstetter or the 9/11 panel on connecting the dots. That privileging of collection also runs through many common analytic techniques. In analysis of competing alternatives, for instance, the competition among the hypotheses turns on evidence. To be fair, the technique's creator, Richards Heuer, wisely notes that "the most probable hypothesis is probably the one with the least evidence against it, not the most evidence for it."[13] Yet evidence—collection—is still the metric. Worse, since adversaries will try to hide critical information, intelligence is least likely to have the information that it most needs.

[13] Heuer, "Psychology of Intelligence Analysis," p. 108.

The emphasis on collection in the cycle has a least three negative consequences. First, it probably leads to too much collection, which is expensive, and at the expense of too little analysis, which is cheap.[14] Reviews of finished products almost always include some analysis of "gaps"—where was intelligence missing and thus where might different or better collection have helped "fill the gap"? That is exacerbated by the requirements process, which provides a virtually bottomless vessel to fill with information. This emphasis on quantity is explicit in the work of nineteenth-century German historiographer Ernst Bernheim, to whose work Kent often referred.[15] For Bernheim, it was exactly the *quantity* of data that could make the *Auffassung*—comprehension of the true meaning of the evidence—objective. "For true certainty . . . the available data must be so abundant and dense that it only allows room for a single connection." When data are insufficient, then "several combinations will be probable in the same degree," allowing the historian to make only "a hypothesis"—a condition which the historian should remedy by making every effort to get more data.[16] Yet review after review suggests that intelligence collects more than it can absorb. In the 1970s, for instance, James Schlesinger, at the time head of the Office of Management and Budget (OMB), criticized the US intelligence community for the "strong presumption . . . that additional data collection rather than improved analysis will provide the answer to particular intelligence problems." [17]

The second problem with emphasizing collection is that doing so also emphasizes what can be collected. Intelligence has often been accused of committing errors akin to the famous "looking for the keys under the lamppost." Thirty years ago, one of us (Agrell) wrote of technical intelligence that it made it possible for the most advanced countries to get "an almost complete picture of the strength, deployment and activity of foreign military forces." The negative was an over-emphasis on what could be counted. Intelligence became "concentrated on evaluation and comparison of military strength based exclusively on numerical factors."[18] Agrell was emphasizing the intangibles even

---

[14] Anthony Olcott, "Stop Collecting – Start Searching (Learning the Lessons of Competitor Intelligence)," unpublished paper.

[15] The book Kent cites, *Lehrbuch der historischen Methode und der Geschichtsphilosophie,* was first published in 1889 and was subsequently republished in five further editions, the last of which appeared in 1908—the one to which Kent refers.

[16] Patrick Kelly, "The Methodology of Ernst Bernheim and Legal History," as quoted in Olcott, "Stop Collecting–Start Searching."

[17] "A Review of the Intelligence Community" (Washington DC: Office of Management and Budget 1975), pp. 10, 11.

[18] Wilhelm Agrell, "Beyond Cloak and Dagger," in *Clio Goes Spying: Eight Essays on the History of Intelligence,* ed. W. Agrell and B. Huldt (Lund: Lund Studies in International History 1983), pp. 184–185.

of military strength, like morale, determination, leadership, and the like. The broader point is that it is easier to collect data about what is going on than what a target is thinking or what will happen.

That point is driven home by recent US experiences with intelligence in support of warfighters. While, tactically, those DIMPIs are welcome, they are not enough even on the battlefield. One critique emphasized that a requirements-driven process means that intelligence spends most of times trying to fill holes rather than questioning assumptions or exploring new hypotheses.[19] Intelligence may begin by shaping leaders' interest, when there is little information, but quickly turns to feeding that interest, and intelligence becomes more and more bound to initial assumptions. The process develops old insights and fails to notice the changing environment. There is more and more focus on executing "our" plan while neglecting information that casts doubt on the assumptions undergirding that plan. In principle, long-term assessments should play that role, but they are seldom done given the urgency of the immediate, usually defined in very operational terms.

Another critique echoed those themes; it was of the war in Afghanistan, and one of its authors was in charge of intelligence for the coalition.[20] In principle, strategic and tactical overlapped; but in practice the preoccupation with the threat from improvised explosive devices (IEDs) meant that "the tendency to overemphasize detailed information about the enemy at the expense of the political, economic and cultural environment [became] even more pronounced at the brigade and Regional Command levels."[21] Those latter levels had lots of analysts, and so did the Washington-based agencies, but those analysts were starved for information from the field.

By contrast, there was lots of information on the ground, in the heads of individual soldiers, Provincial Reconstruction Teams (PRTs), and aid workers. The challenge was getting those data shared upward. The critique recommended that the higher levels of command send analysts to the units on the ground, noting that the task would not be so hard, for already there were plenty of helicopters shuttling between PRTs and brigade and battalion headquarters. Moreover, such broader analysis as did get done was very stovepiped, addressing governance or narcotics or another topic. It was like a sportswriter discussing all the goalkeepers in a league without talking about the teams. The authors found only one example of "white" analysis, not just "red"—that is,

---

[19] Steven W. Peterson, *US Intelligence Support to Decision Making* (Cambridge, MA: Weatherhead Center for International Affairs, 2009).

[20] Michael Flynn, *Fixing Intel: A Blueprint for Making Intelligence Relevant in Afghanistan* (Washingnton, DC: Center for a New American Century, 2010).

[21] Ibid., p. 8.

assessments not just of the bombers but of the circumstances that produced them—about Kandahar province done by a Canadian agency.[22]

The critique's principal recommendation is also provocative for intelligence and its connection to policy. The military's existing fusion centers in Afghanistan were fine, but they operated at the level of SCI—secret, compartmented intelligence. Yet most of the "white" analysis was open source. Thus, those fusion centers might have been complemented with Stability Operations Information Centers. Those would have been places for Afghan leaders to come and share information, along with members of the NATO international coalition, ISAF (International Security Assistance Force).

The third problem with privileging collection is more subtle but probably also more powerful. The collect-then-analyze model is driven by requirements (and, in the United States by the National Intelligence Priorities Framework, or NIPF). Collection then functions like a web scraper, scouring the world for information against those requirements and storing it for possible use, now or later. On first blush, that sounds impressively Google-esque. Yet, as Anthony Olcott notes, competing search engines produce the same results as little as 3 percent of the time.[23] They usually produce very different results, and even if the same websites appear, the search engines frequently rank them differently.[24] And that leaves aside the so-called dark web, which does not show up in searches but is perhaps a thousand times larger than the known web.[25] As a result, "collection as Google" is certain to omit far more relevant information than it gathers.

Confirmation Bias and Its Kin

So, collection inevitably is selective. A study one of us (Treverton) did described the analytic trade craft of US intelligence during the 1970s and 1980s. That study observed: "For all the centers and task forces, analysts still mostly work alone or in small groups. Their use of formal analytic methods, let alone computer-aided search engines or data-mining, is limited. Their basis for analysis is their own experience, and their tendency is to look for

---

[22] *District Assessment: Kandahar City, Kandahar Province* (Ottawa: Canadian Department of Foreign Affairs and International Trade, 2009), cited in Ibid., p. 18.

[23] Amanda Spink et al., "Overlap among Major Web Search Engines," in *Third International Conference on Information Technology: New Generations* (Los Alamitos, CA: Institute of Electrical and Electronics Engineers, 2006).

[24] Anthony Olcott, "Institutions and Information: The Challenge of the Six Vs" (Washington, DC: Institute for the Study of Diplomacy, Georgetown University, 2010).

[25] Chris Sherman and Gary Price, *The Invisible Web: Uncovering Information Sources Search Engines Can't See* (Meford, NJ: Information Today, 2001).

information that will validate their expectations or previous conclusions."[26] There was no intention to caricature, and indeed the practice made a certain sense at the time, against a primary target, the Soviet Union, that was ponderous. The broader point, though, is that both collectors and analysts select the information they will save or employ. The collect-then-analyze procedure, plus requirements and still more taskings, only reinforces that selection.

Such selection raises the specter of confirmation bias, the human tendency to select or interpret information in a way that confirms pre-existing hypotheses.[27] The world is rife with examples, across many of the domains surveyed in this book.[28] Like Garrison Keillor's Lake Wobegon where all the children are above average in intelligence, 93 percent of US drivers think they are better than average. Studies have found 68 percent of university faculty members thought they were in the top quarter in teaching performance, while a quarter of students thought they were in the top 1 percent in leadership ability.[29] Across fields, experts tend to be overconfident in their estimates, sometimes wildly so. One study of doctors found that while they thought they were 80 percent certain in diagnosing pneumonia, they actually were correct only 18 percent of the time. In law enforcement, studies have shown that investigators and interrogators are confident they can tell when a suspect is lying when, in fact, they are no more accurate than untrained college students. And, to boot, the cues in behavior they looked for were, in fact, not correlated with lying.

Even more disheartening, given selection, more information tends to make the disparity between confidence and accuracy worse, not better. New information increases confidence more than accuracy. Moreover, not only do experts tend to look for information that will confirm their methods and their conclusions but they remember their forecasts as better than they were. Philip Tetlock found that experts' memories conveniently moved their previous estimates 10 percent to 15 percent closer to accurate.[30] That is, if they predicted an event would occur with a probability of 60 percent and it did in fact occur, they remembered their confidence as 70 percent or more. Conversely,

[26] Gregory F. Treverton and C. Bryan Gabbard, *Assessing the Tradecraft of Intelligence Analysis* (Santa Monica: RAND Corporation, 2008), pp. 33–34.

[27] For a nice discussion of confirmation bias and what might be done about it, see Paul Lehner, Avra Michelson, and Leonard Adelman, *Measuring the Forecast Accuracy of Intelligence Products* (Washington DC: MITRE Corporation, 2010).

[28] Raymond S. Nickerson, "Confirmation Bias: A Ubiquitous Phenomenon in Many Guises," *Review of General Psychology* 2, no. 2 (1998): 175–220.

[29] Lehner, Michelson, and Adelman, "Measuring the Forecast Accuracy of Intelligence Products," pp. 4–5.

[30] Philip Tetlock, *Expert Political Judgment: How Good Is It? How Can We Know?* (Princeton, NJ: Princeton University Press, 2005), p. 149.

if the predicted event didn't occur, they recalled their confidence as only 50 percent.[31]

Intelligence analysts may be particularly prone to these failings because of the verbal imprecision with which their estimates are all too often stated. It is easy for memory to inflate "a fair chance" into near certainty if the forecast event actually happens. "May occur" may be remembered as accurate no matter which way the event actually turned out. At least since Sherman Kent, those who manage intelligence analysis have tried to introduce more precision into the language of estimates. In his charming account, his effort to quantify what were essentially qualitative judgments, what he called the "mathematician's approach," met opposition from his colleagues, ones he labeled the "poets."[32] Kent regarded them as defeatist. They saw his effort as spurious precision in human communications. In the latest US effort, American National Intelligence Estimates now come with a paragraph defining terms and a chart arranging those words in order of probability.[33]

A close kin of the confirmation bias is common sense and circular reasoning. As Duncan Watts, who moved from biology to sociology, puts it: "common sense often works just like mythology. By providing ready explanations for whatever particular circumstances the world throws at us, commonsense explanations give us the confidence to navigate from day to day and relieve us of the burden of worrying about whether what we think we know is really true."[34] Common sense is durable. Perhaps the most famous episode of that durability is the US September 1962 National Intelligence Estimate that the Soviet Union would not install nuclear missiles in Cuba, mentioned in the previous chapter. That was perhaps the most prominent mistake Sherman Kent's estimators made. Yet in his postmortem, Kent attributed the mistake more to Khrushchev than to the estimators. In Kent's words, "We missed the Soviet decision to put the missiles into Cuba because we could not believe that Khrushchev could make a mistake."[35] Such was Kent's faith in what he called "the dictates of the mind."[36] What Khrushchev did was not rational.

---

[31] For a psychological analysis for the roots of this hindsight bias, see Neal J. Roese and Kathleen D. Vohs, "Hindsight Bias," *Perspectives on Psychological Science* 7, no. 5 (2012): 411–426.

[32] See Sherman Kent, "Words of Estimative Probability," *Studies in Intelligence*. 8, no. 4 (Fall 1964): 49–65.

[33] See, for instance, the declassified Key Judgments in National Intelligence Council, *Iran: Nuclear Intentions and Capabilities*, National Intelligence Estimate (Washington, DC, 2007).

[34] Duncan J. Watts, *Everything Is Obvious: Once You Know the Answer* (New York: Crown, 2011), p. 28.

[35] Sherman Kent, "A Crucial Estimate Relived," *Studies in Intelligence* 8, no. 2 (1964).

[36] Sherman Kent, *Writing History* (New York: F.S. Crofts, 1941), p. 4.

Common sense is thus close to mirror imaging: what would make sense for us to do if we were in their shoes? To be fair, even in retrospect, it is hard to comprehend Khrushchev's decision. It simply seems to carry too much risk for too little potential gain, all the more so from a conservative leader like him. Yet there is a certain circular reasoning to all these cases. As Olcott puts it: "we explain phenomena by abstracting what we consider to be causal factors from the phenomena, and then attribute their success to the presence of those characteristics, in effect saying "X succeeded because it had all the characteristics of being X." Even more important, when a "common sense" explanation fails, we tend to generate explanations for that failure much as Kent and his team sought explanations for their miscalculation in the NIE, in effect arguing that "Y failed because it did *not* have the characteristics of being X. If only the situation had included *a, b,* and *c,* then it would have been a success."[37]

One source of mistaken assessment lies more on the shoulders of policymakers than intelligence analysts. That is the "inconvenient" alternative. A vivid example is the imposition of martial law in Poland by the Polish government in 1981. That was probably not the "worst" alternative from NATO's perspective; Soviet intervention might have been worse in many senses. Yet all of US and NATO planning was based on the assumption that if martial law came to Poland, it would be imposed by the Soviet Union. For the Poles to do it themselves was inconvenient. When one of us (Treverton) teaches about the intelligence-policy interaction, one of his cookbook questions is this: ask yourself not just what is the worst alternative, but also what is the most inconvenient alternative?

All too often the inconvenient alternative is specifically inconvenient to the plans of a particular government or leader. On July 1, 1968, President Johnson announced, at the signing of the Non-Proliferation Treaty, that the Soviet Union had agreed to begin discussions aimed both at limiting defenses against ballistic missiles and reducing those missiles. But neither date nor place for the talks had been announced when, on August 20, the Soviet Union began its invasion of Czechoslovakia, an event that postponed the talks and denied the president a foreign policy success. Intelligence presaging the Soviet invasion of Afghanistan was similarly inconvenient, hence unwelcome, for President Jimmy Carter in 1979. First indications came when Carter was on the plane to Vienna to sign the SALT II treaty. Later, when the evidence of the invasion was clearer, SALT II was awaiting ratification on Capitol Hill. In both cases, the warning was unwelcome to the president.[38]

---

[37] Olcott, "Stop Collecting–Start Searching."

[38] For the US assessments on Afghanistan, see D. J. MacEachin, *Predicting the Soviet Invasion of Afghanistan: The Intelligence Community's Record* (Washington, DC: Center for the Study of Intelligence, Central Intelligence Agency, 2002).

All of these kindred failings—mirror imaging, excluding the inconvenient alternative, circular reasoning—rest on what makes sense, either to them as we understand them or to us as we imagine ourselves in their shoes. Robert Jervis concluded his postmortem on the fall of the Shah of Iran by observing that intelligence analysts simply found it "inconceivable that anything as retrograde as religion, especially fundamentalist religion, could be crucial."[39] It didn't make sense. Similarly, just as common sense cut off Kent's estimators from asking what might move Khrushchev to take the risk of deploying nuclear missiles in Cuba, so, too, it stayed later analysts from asking why Saddam Hussein might pretend he had weapons of mass destruction (WMD) in 2002 that he didn't have, or didn't know he didn't have. Neither line of reasoning made sense. What information had been collected was used selectively to confirm the wisdom. In the Iraq case, that collection, primarily the United Nations (UN) inspections, was simply dismissed despite the fact that, in a Bayesian sense, the more the inspectors *didn't* find weapons of mass destruction, the more seriously policymakers ought to have taken the possibility that he didn't in fact have them.

Most of what get labeled "intelligence failures" result in fact from the interaction of intelligence and policy. When the two come to share a common mindset around a common story or "concept," then the interaction becomes a feedback loop in which questions that cut against the concept are not likely to be asked and disconfirming evidence not likely to be offered as an answer. The supposed presence of Iraq WMD is the poster child recent case, and the 1973 Arab-Israeli war is the most celebrated (and most studied). In 2002 the concept was that Saddam Hussein *must have WMD*; after all, he had them before and he had reason to have them later. Not only did Secretary of Defense Donald Rumsfeld's famous quote—absence of evidence is not evidence of absence—do brutality to Bayesian reasoning; it was also graphic testimony to the power of the story or concept.

In the case of the 1973 war, the concept was more complicated but its core was that Egypt's Anwar Sadat would not start a war he could not win. The concept was then elaborated into equipment or capabilities he needed to have a better chance of winning—especially longer-range fighter-bombers and better surface-to-surface missiles to dent Israeli air superiority. The absence of those capabilities then became indicators that an attack was not forthcoming. In the WMD case, the evidence was scanty—that "absence" again—a fact that was explained away by deception or good hiding or having moved the capability. The 1973 war is more striking because the Israelis had lots of evidence of a possible impending attack,

---

[39] Jervis, *Why Intelligence Fails.*

including from Jordan's King Hussein. They also, it turned out, had a spy close to Sadat's inner circle. For most of the runup to the war, his detailed reporting on capabilities tended to play into the concept. He reported in detail on what the Soviets were—or, rather, weren't—providing by way of capabilities to Egypt, and so reinforced the idea that Sadat wouldn't start a war he couldn't win.[40]

Roads Not Taken

Traditional intelligence analysis, especially in the United States, took "us" out of the analysis. It was "foreign" intelligence about "them" over there, not us. Until September 11, 2001, and especially during the Cold War, that separation of home and abroad could be justified in a fashion. As former US secretary of defense Harold Brown quipped about the US-Soviet nuclear competition, "When we build, they build. When we stop, they build."[41] While various countries, especially the United States, hoped that their policies would influence Moscow, as a first approximation intelligence analysts could presume that they would not. The Soviet Union would do what it would do. The challenge, in the first instance, was figuring out its likely course, not calibrating influence that other nations might have over that course.

The roots of that separation in the history of intelligence are suggestive of how big a change will be necessary to undo it. Much of the legacy of US intelligence analysis, and by extension that of its friends and allies, owes to Sherman Kent, the Yale University history professor who had served in the wartime Office of Strategic Services, then returned to the Central Intelligence Agency (CIA) in 1952 to help found, then run, the new Office of National Estimates. Back at Yale, Kent wrote *Strategic Intelligence for American World Policy,* and he is often regarded as the father of modern intelligence analysis.[42] His view did powerfully shape, and continues to shape, how the craft of intelligence analysis is performed. As sometimes happens, the power of Sherman Kent's view of intelligence, for not just the United States but also its friends and allies as well,

---

[40] U. Bar-Joseph, *The Watchman Fell Asleep: The Surprise of Yom Kippur and Its Sources* (Albany: State University of New York Press, 2005), pp. 21–23. For the employment of and reliance on "defining factors" in intelligence assessments, see Wilhelm Agrell, *Essence of Assessment: Methods and Problems of Intelligence Analysis* (Stockholm: National Defence College, Center for Asymmetric Threat Studies, 2012), Chapter 7.

[41] "Respectfully Quoted: A Dictionary of Quotations," ed. Suzy Platt (Washington, DC: Library of Congress, 1989), p. 80.

[42] Kent, *Strategic Intelligence for American World Policy.*

is most vividly seen through a famous critique by Willmoore Kendall, which suggests the outlines of a "road not taken."[43]

Some of Kendall's review builds on Kent's words to make observations that look eerily prescient from the perspective of more than a half century later. For instance, Kent had feared that the security requirements of covert collection would make overt collection difficult. This may be particularly so for the United States, which chose to combine clandestine collection and intelligence analysis in one agency, the CIA. It is hard to overstate how much the needs of the clandestine service determine how the agency is run. When one of us (Treverton) sat in the CIA building running the process of producing National Intelligence Estimates, his schedule was dutifully typed and retyped several times a day as the calendar changed. But it was always classified "Secret." The reason was that he might meet with a clandestine service officer under cover. It was indicative that rather than following the usual US intelligence practice of lopping surnames, or even leaving a blank, the appointment appeared in true name and so required the schedule to be classified even if no such meeting was on the calendar. Treverton recalled a conversation with a friend who had been director of the CIA's directorate of analysis. He said, tellingly, "It looks on the organization chart like the director of intelligence and of operations [the clandestine service] are equals. But they're not. For some purposes, I had to get clearance from DO [directorate of operations] officers several levels below the director."

Kendall went further to observe that "our present intelligence arrangements . . . enormously exaggerate the importance to covert collection, and yet permit it to yield shockingly small dividends."[44] This book is not about clandestine collection or espionage, and it hazards no opinion about whether Kendall's view holds more than a half century later. Yet a serious review of espionage is long overdue, at least in the United States. The review would have to be classified, but some results could be public. And, in any event, it would need to include a range of people beyond intelligence professionals and especially beyond the clandestine service itself. From what is publicly known, the record is not impressive. The most important Soviet spies for the United States were "walk-ins"—that is, recruited themselves. All US agents in Cuba were double agents, really working for Cuba. And there has been a cottage industry of memoirs emphasizing that the incentives in the service encourage numbers

---

[43] Willmoore Kendall, "The Function of Intelligence," *World Politics* 1, no. 4 (1949): 542–552.

[44] Ibid., p. 545. See Anthony Olcott, "Revisiting the Legacy: Sherman Kent, Willmoore Kendall, and George Pettee—Strategic Intelligence in the Digital Age," *Studies in Intelligence* 53, no. 2 (2009): 21–32.

of recruited spies, which can be evaluated immediately, rather than what really matters—the quality of their information, which may take years to assess.[45]

In analysis, the first stretch of the road not taken is the separation of "them" from "us"—the sense that if politics stopped at the water's edge going outward, intelligence should stop at the water's edge coming inward. The separation led Kent—and still leads intelligence analysis—to draw a bright white line separating foreign, other states, and what Kent called "our own domestic scene." As a result, in Kendall's words, "intelligence reports . . . never, never take cognizance of United States policies alternative to the one actually in effect, such problems being "domestic" matters."[46]

In his book, Kent endorsed Walter Lippmann's view a quarter century earlier that "every democrat feels in his bones that dangerous crises are incompatible with democracy, because the inertia of the masses is such that a very few must act quickly."[47] Thus, in Lippmann's words, "The only institutional safeguard is to separate, as absolutely as it is possible to do so, the staff which executes from the staff which investigates."[48] In those circumstances, the only way to ensure what Kent called "impartial and objective analysis" was to create, in Lippmann's words, "intelligence officials" who would be "independent both of the congressional committees dealing with that department and of the secretary at the head of it" so that "they should not be entangled either in decision or in action."[49] For Kendall, by contrast, intelligence and policy together needed to confront what Kendall called "the big job—the carving out of the United States destiny in the world as a whole."[50]

The rub with the Kent view, of course, is that his "impartial and objective analysis" might also be irrelevant. With such distance between the two, intelligence had little hope of knowing what policy officials knew or needed to know, on what timetable and in what way. Kent himself recognized the problem in a 1948 letter to then-Director of Central Intelligence Admiral Roscoe Hillenkoetter: "Since [ORE, the CIA's Office of Research and Analysis] has no direct policy, planning, or operating consumer to service within its own organization . . . it is likely to suffer . . . from a want of close, confidential, and friendly guidance." His solution was that

[45] The exaggerated and in many ways dysfunctional role of clandestine collection is not unique for the United States but was widespread during the Cold War—for instance, in the German BND under the directorship of Gerhard Gehlen, see Jeffery T. Richelson, A Century of Spies: Intelligence in the Twentieth Century (New York: Oxford University Press, 1995).

[46] Kendall, "The Function of Intelligence," p. 549.

[47] Walter Lippmann, Public Opinion (New York: Harcourt, Brace, 1922), p. 272.

[48] Ibid., pp. 384, 386.

[49] Kent, Strategic Intelligence for American World Policy, p. 100. Ibid., p. 61.

[50] Kendall, "The Function of Intelligence," p. 548.

ORE should be brought into closest and most direct contact with consumers such as the National Security Council ... having an ORE officer represent CIA (or participate in CIA's representation) at NSC staff discussions would have two great benefits: (a) It would assure ORE of knowing the precise nature of the consumer's requirements; and (b) it would enable ORE to convey to the consumer the precise dimensions of its capabilities. It is to be noted that these two matters interlock: when the consumer knows ORE's capabilities, he may change the dimensions of this requirement (add to it, lessen it, or reorient it), and, when ORE knows the precise dimensions of the requirement, it may deploy its resources in such a fashion as to enlarge its capabilities. So long as liaison between consumer and ORE is maintained by someone not possessed of the highest professional competence in matters of substance and firsthand knowledge of ORE's resources, that liaison is almost certain to be inadequate for the purposes of both ORE and the consumer.[51]

The idea was hardly new. Several years earlier, Assistant Secretary of State Donald Russell had tried something very similar, a recommendation made by the then-Budget Bureau (later OMB, the Office of Management and Budget), which was a participant in postwar discussions of intelligence. For Russell, "the principal intelligence operations of the Government should be organized at the point where decision is made or action taken, i.e., at the departmental, or lower, level and not within any single central agency." If not, "the policy recommendations of a research unit which is not organizationally integrated with operations are very likely to be theoretical judgments with little basis in reality."[52] Yet Russell's initiative died. The creation of the CIA was precisely that single central agency against which Russell had warned. The logic of the CIA's creation was Pearl Harbor and a check on department intelligence: it should make sure it has all the information and so can serve as a counterweight to temptations by departmental intelligence agencies to cut their assessments to suit the cloth of their operators.

The Russell idea was not revived despite a series of blue-ribbon panels, mostly addressing efficiency but all making the point that much of the inefficiency derived from the lack of feedback from policy officials. In 1966, the CIA inspector general—in what is generally referred to as the Cunningham Report—responded to criticisms that the community had failed to "adequately

---

[51] *Foreign Relations of the United States, 1945–1950: Emergence of the Intelligence Establishment* (Washington, DC: United States Department of State, 1996).

[52] Ibid., document 81, November 3, 1945; and document 82, December 29, 1945.

consider the broader question of the slowly developing Sino-Soviet dispute." The report, though, concluded that the CIA collected "too much information and that, failing to get important information, it was flooding the system with secondary material," thus "degrading production, making recognition of significant information more difficult in the mass of the trivial." The reason for this was telling: "there was no definition of what the government really needed from intelligence, so the community operated on its own assumptions, which tended to cover everything, just in case."[53]

A few years later, James Schlesinger, then at OMB, conducted a review of the community.[54] He worried that "the impressive rise in [the] size and cost" of intelligence had not produced "a commensurate improvement in the scope and overall quality of intelligence products." His reason was the same as five years earlier, that "the consumer frequently fails to specify his product needs for the producer; the producer, uncertain about eventual demands, encourages the collector to provide data without selectivity or priority; and the collector emphasizes quantity rather than quality."

Another five years later, the Church Committee again argued that "collection guides production rather than vice-versa."[55] And again, the reason for this "glut of paper" was that

> evaluation of the intelligence product by the consumers themselves is virtually non-existent . . . . Rarely, if ever, do high officials take the time to review the product carefully with the analysts and explain to them how the product could be improved and made more useful to the policymakers. The intelligence community, then, by default, evaluates its own performance without the benefit of any real feedback.

The second stretch of Kendall's road not taken is based on analysis of Kent's underlying approach. Kent argued that intelligence was the same in peace as in war, but for Kendall the two were very different. Wartime intelligence is

---

[53] The *Cunningham Report*, presented in December 1966, had not been fully declassified. Lines from it are quoted in the Church Committee Report, "Foreign and Military Intelligence: Book 1: Final Report of the Select Committee to Study Governmental Operations with Respect to Intelligence Activities" (Washington: United States Senate, 1976), hereafter cited as *Church Committee Report*. In addition, the Cunningham Report was quoted and also summarized in "A Historical Review of Studies of the Intelligence Community for the Commission on the Organization of the Government for the Conduct of Foreign Policy," (document TS–206439–74) (1974).

[54] "A Review of the Intelligence Community," pp. 1, 9, prepared by the Office of Management and Budget under the direction of OMB Deputy Director James Schlesinger, March 10, 1971.

[55] *Church Committee Report*, p. 277.

primarily tactical because the enemy is known and the objectives are clear. By comparison, peace requires more strategic analysis because neither a nation's objectives nor those of adversaries can be taken as a given. The "big job" had to be defined; it could not simply be assumed. Moreover, the emphasis on wartime also underpins the distinction between us and them. That puts the assessing of "them" "in the hand of a distinct group of officials whose "research" must stop short at the three-mile limit even when the threat they are following runs right across it, and yet which tells itself it is using the scientific method."[56]

Perhaps the Cold War was enough like a hot one to make the conflation of war and peace less dangerous than it is now. As Olcott and Kerbel put it: "The kind of analytic support that Kent envisioned—analysts standing behind policymakers 'with the book opened at the right page, to call their attention to the stubborn fact they may neglect' almost inevitably drives analytic support toward tactical intelligence, rather than the strategic, but it worked well for the IC's [intelligence community's] Cold War glory years, because the nature of the Soviet Union and the means to face it were such that tactics all but merged with strategy."[57] Notice that the great successes of Cold War intelligence were puzzles—do Soviet missiles have multiple warheads, how accurate are they? They weren't exactly tactical, but they were technical and about capabilities. Soviet strategy was more assumed than analyzed.

Kent's emphasis on professionals, on both the policy and intelligence side, and on "producers" and "consumers" was, for Kendall, a quintessentially bureaucratic perspective. It made intelligence analysts "mere research assistants to the George Kennans." More tellingly, it excluded elected officials and what Kendall thought most crucial, "communication to the *politically* responsible laymen of the knowledge which . . . determines the 'pictures' they have in their heads of the world to which their decisions relate." For Kendall, Kent's approach reinforced a "crassly empirical conception of the research process in the social sciences," one organized by region and dominated by regional specialists. In the terms of this chapter, it was more puzzle-solving *than* framing mysteries. The task was "somehow keeping one's head above water in a tidal wave of documents, whose factual content must be "processed," and the goal was prediction, "necessarily understood as a matter of projecting discernible empirical trends into an indefinite future." Again, the approach may have worked well enough during the Cold War. The Soviet Union was linear and static enough that it could be analyzed tactically. It was also complicated

---

[56] Kendall, "The Function of Intelligence," p. 549.

[57] Josh Kerbel and Anthony Olcott, "Synthesizing with Clients, Not Analyzing for Customers," *Studies in Intelligence* 54, no. 4 (2010): 13. The Kent quote is from *Strategic Intelligence for American World Policy*, p. 182.

enough that Kent's style of analysis—breaking down or reducing, from the Greek root—also worked tolerably well.[58] By contrast, Kendall argued that the research process should be open to "'theory' as it is understood in economics and sociology," and enable analysts to "work under conditions calculated to encourage thought." Despite the many references in Kent's book to social scientists and social science, he "never employs in that connection the words *theory* and *theorist*."[59]

The road not taken might have given more pride of place to strategic intelligence and to unclassified collection. It might have explicitly acknowledged that "us" and our actions cannot be excluded from the analysis—and that the exclusion is all the more damaging the more important an issue is to the United States and the more active it is with respect to it. Kent bears less responsibility for the "bright white line" that has divided intelligence from policy, especially in the United States—that was as much a reaction to the Pearl Harbor failure—but his bureaucratic view wasn't inconsistent with that separation. So the road not taken might have included more interaction between the two, especially at the more "political" levels of the government, and including Congress. That road not taken might also have sought to open space for if not theory, then at least thought.

A final hint of the road not taken comes from another academic turned wartime intelligence professional writing at the same time, George Pettee. Like Kent, he conceived intelligence analysis primarily as a research task, but he went further than Kendall in arguing that it made no sense if not connected to policy: "The failure to define clearly the special province of 'strategic' or 'national policy' intelligence . . . meant in the past that the conduct of the work lacked all the attributes which only a clear sense of purpose can give."[60] But his major contribution to the road not taken was his emphasis on what today might be called "cultural intelligence." Wartime intelligence had gotten the numbers right; what it got wrong were the contexts and assumptions in which analysts embedded those numbers. With Kendall, Pettee argued that beliefs exist separately from data. What intelligence too often missed reflected "a long record of American ignorance or misunderstanding of what makes world politics operate."[61]

[58] Olcott, "Revisiting the Legacy."

[59] Kendall, "The Function of Intelligence," pp. 549–551.

[60] George Pettee, *The Future of American Secret Intelligence* (Washington, DC: Infantry Journal Press, 1946), p. 65.

[61] Ibid., p. 39.

4

Intelligence Challenges
in Scientific Research

Scientific research, unlike intelligence analysis, is not immediately identified with a string of recurring failures. True enough, scientific failures happen from time to time and might attract considerable public attention. But the vast difference is the predominance of *success*. Commissions investigating major scientific failures are rare,[1] while scientific promotion is ever present and the Nobel prizes are awarded annually for scientific and cultural achievements. When did a brilliant intelligence analyst receive the corresponding international recognition and publicity? Or as John Keegan remarks in his book on the development of military intelligence from Napoleon to Al Qaida: it is difficult enough, in any case, to make a reputation as a staff officer, but while there are a number of celebrated operations officers and chiefs of staff, there are almost no famous intelligence officers.[2] So those who care for career opportunities should put their bet on academia rather than intelligence. No wonder then that the imagined round table of *Intelligence Minds* never has come to be.

---

[1] The main aspect of scientific under-performance that has resulted in reactions similar to intelligence postmortems has been research frauds, perceived as a major threat to public and collegial credibility. There is considerable research on scientific fraud and misconduct. See, for example, Stephan Lock, *Fraud and Misconduct in Medical Research*, ed. Frank Wells (London: BMJ Publishing Group, 1993); Marcel C. LaFollette, *Stealing into Print: Fraud, Plagiarism, and Misconduct in Scientific Publishing* (Berkeley: University of California Press, 1992); David J. Miller and Michael Hersen, *Research Fraud in the Behavioral and Biomedical Sciences* (New York: John Wiley, 1992); Sheldon Krimsky, *Science in the Private Interest: Has the Lure of Profits Corrupted Biomedical Research?* (Lanham: Rowman and Littlefield, 2003); Stephan Lock and Frank Wells, *Fraud and Misconduct: In Biomedical Research*, ed. Michael Farthing (London: BMJ Books, 2001).

[2] J. Keegan, *Intelligence in War: Knowledge of the Enemy from Napoleon to Al-Qaeda* (Knopf Doubleday, 2003), p. 384.

"There's No Success like Failure ..."

The difference between the two domains is not mainly one of career opportunities and the prospects of reward or scorn. In these two, the very concept
of success and failure is different.[3] As noted in the previous chapters, all perceived failures in which intelligence in some way was engaged tend to be
regarded as *intelligence* failures. While these assumed intelligence failures do
not always reach the public domain, they are under certain conditions elevated
into major political issues and investigated by commissions, parliamentarians,
or public prosecutors (see Chapters 7 and 8). Those investigations seldom are
conducted by intelligence experts—who may indeed be regarded as part of the
problem—yet the chances that those scrutinized will emerge unscathed at the
far end nevertheless are slim.[4] In science, the dividing line between success
and failure is far from unambiguous. Scientific failures could have disastrous
consequences, depending on how the results are interpreted and implemented
by external actors. But scientific failures, in the form of inaccurate or faulty
assumptions, could also be important steps in the cumulative scientific process. This is the central element in Karl Poppers's classical essay on the philosophy of science, *Conjectures and Refutations*. Popper defines what constitutes a
scientific statement and, as a consequence, scientific success: "Only if a theory
successfully withstands the pressure of these attempted refutations can we
claim that it is confirmed or corroborated by experience."[5]

Science, thus, rests on a fundament of successive failed refutations—that is,
observations, experiments, or logical reasoning that in the end simply did not
hold water—leaving the original hypothesis increasingly validated by every
failed serious attempt to refute it. Without these failed and, as it turned out,
misconceived hypotheses, science as defined by Popper would degenerate not
only into mediocrity but also into metaphysics, a belief system not based on
verifiable observations and refutable reasoning. Transferred from the intelligence domain to the realm of science, the Iraqi WMD estimates would have
constituted a classic case of upholding a paradigm within normal science and
the attempts to construct supporting hypotheses to explain anomalies—such

[3] The line "there's no success like failure" is quote from Bob Dylan, "Love minus zero/no
limit," on *Subterranean Homesick Blues* (1965). The verse nevertheless continues: "... and failure's
no success at all."

[4] Several of the postmortems of the estimates on Iraqi WMD prior to the 2003 war were conducted from a political and judicial rather than an intelligence perspective. True enough, intelligence organizations in some cases prefer amateur investigations that could be more susceptible
to defensive deception and coverup for irregularities.

[5] Karl Raimund Popper, *Conjectures and Refutations: The Growth of Scientific Knowledge*
(London: Routledge and Kegan Paul, 1969), s. 256.

as the failure of the UNMOVIC (United Nations Monitoring, Verification and Inspection Commission) inspectors to find any evidence at suspected sites. The end result, that the WMD simply were not there, would perhaps have been a bit embarrassing for some of those involved but not a failure for science; on the contrary, refutation had prevailed.

Quality checking, the establishment of whether a reported result meets established scientific standards, is generally conducted through peer review, in which a collegial body screens and assesses the scientific basis, validity, and originality of a manuscript, a research project, or a research group. The peer review system reflects a fundamental difference in the professional status and not least autonomy of science compared to intelligence analysis. It is hardly conceivable that a nonscientific entity would be tasked to screen and evaluate scientific findings, at least not in a Western society where science is not subordinate to ideological or religious metaphysics of the kind Popper turned against in his writings.[6] Peer review can be seen as an effort to institutionalize the self-correcting principles of science. A flawed observation or a misconception will sooner or later be scrutinized by others or confronted with other incongruent findings. Misconceptions are therefore neither damaging in the long run nor surprising. As Zhores Medvedev writes in the concluding chapter of his famous account of the Lysenko affair (discussed later in the chapter): "In science the appearance of a false doctrine is a natural process: they are the extreme variants of essential hypotheses, assumptions, and theories. The majority of hypotheses, by the very nature of things, must be incorrect and incomplete reflections of actual phenomena."[7]

In the scientific domain it would, as Medvedev underlines, be absurd to criticize a scientist who advances a hypothesis for discussion and examination, which eventually turns out to be wrong.[8] Scientific controversies are thus not a sudden malfunction of the system but a necessary and permanent precondition for its survival. A high degree of lasting scientific unity could, on the contrary, be regarded as an indication of stagnation and crisis, of the hegemony of a given paradigm where the scientific community develops an orthodoxy, undermining the self-correcting mechanisms and perverting peer review into a status-quo preserving institution of extended groupthink.[9]

---

[6] Karl Raimund Popper, *The Open Society and Its Enemies*, Vol. 1: *The Spell of Plato* (London: Routledge and Kegan Paul, 1945); Karl Raimund Popper, *Open Society and Its Enemies*, Vol. 2: *The High Tide of Prophecy; Hegel, Marx and the Aftermath* (London: Routledge and Kegan Paul, 1947); Karl Raimund Popper, *The Poverty of Historicism* (London: Routledge and Kegan Paul, 1960).

[7] Z. A. Medvedev and T. D. Lysenko, *The Rise and Fall of T.D. Lysenko* (New York: Columtia University Press, 1969), p. 245.

[8] Ibid.

[9] The classical critique of scientific orthodoxy is formulated by Thomas S. Kuhn, *The Structure of Scientific Revolutions* (Chicago: University of Chicago Press, 1962).

If we compare this with the intelligence domain, the dividing line between science and nonscientific beliefs is far more difficult to distinguish. Not only does intelligence as a proto-discipline lack many of the developed self-correcting mechanisms perceived to be so essential in science, despite a good deal of experimentation in this direction as outlined in Chapter 3,[10] but also the nature of the problems dealt with are often of such a character that it is seldom feasible to conduct intelligence along the line of successive refutations. While perhaps sound in principle, such a practice would bring down disaster on a machinery expected to provide timely delivery of, if not the truth, at least the most accurate and well-founded assessments at hand, not to diverge into an exercise in analytic cleansing and intelligence refutations, however well executed. Moreover, who would actually be the peers in an intelligence context? Robert Jervis observed, when reflecting over the CIA's assessment on Iran in the late 1970s, that there was little peer review in intelligence at this stage. There were certainly reviews, but they were conducted by successive layers of managers and did not constitute an analytic probing.[11]

Controversies in Science

Given these self-correcting mechanisms, scientific controversies are not uncommon and could, as both Popper and Medvedev underline, be regarded as a necessity for the advancement of knowledge. Scientific consensus, therefore, is not necessarily a positive sign, especially not if consensus implies that attempts to refute the dominating theories diminish or cease. In Thomas S. Kuhn's concept of scientific paradigms, however, the established "normal science" develops precisely these conservative and self-confirming mechanisms, where research is accepted if in line with the dominating paradigm, and regarded with suspicion or ignored if not.[12] The actual controversies, however, do not appear in the ideal world of the philosophy of knowledge, but in the real world, one complete with individuals, institutions, schools of thought, and, not least, within an external social and political context.

A scientific controversy can be defined as publicly and persistently maintained conflicting knowledge claims on issues that, in principle, can be determined by

---

[10] The "Devil's Advocate" institution can be seen as a substitute for self-correcting mechanisms within a closed and compartmentalized bureaucracy. For the "Devil's Advocate" method, see in this context Robert Jervis, *Perceptions and Misperceptions in International Politics* (Princeton, NJ: Princeton University Press, 1976), p. 415.

[11] Jervis, *Why Intelligence Fails*, p. 24.

[12] Kuhn, *The Structure of Scientific Revolutions*.

scientific means and where both contenders thus claim scientific authority.[13] This reference to scientific authority helps explain the intensity of many of these controversies. Some scientific controversies are solved, while others are not and remain endemic in academia, sometimes manifesting themselves in the split of disciplines or research fields. Other controversies are simply ignored or embedded in academic diversity; disciplines with distinctly different methods and theoretical foundations normally do not clash since they see each other as too distant to be relevant. Here, no attempted refutation takes place; there is neither much by way of incentives nor a common ground on which a meaningful dialogue could be based.

In other instances, where the contenders are relatively close to each other, share the same, or at least some, theoretical and empirical foundation, and not least advance interpretations with overlapping explanatory claims, controversies cannot normally be ignored or embedded. This is especially the case when these competing claims attract external attention or have more or less immediate relevance in the society. The rift between Scandinavian political science and the emerging peace and conflict research from the 1960s to the 1990s did not stem from differences in the intellectual foundation but rather in different attitudes to values and in subsequent conflicting claims to explain and deal with identical or similar problems, such as the arms race and nuclear deterrence. Some scientific controversies thus resemble conflicts over intelligence puzzles, while others have dimensions similar to mysteries, where the framing of the issue at stake, the key variables, and how they might interact is essential.

However, solving scientific controversies is more complex than just determining whose knowledge claims are justified and whose are not. In controversies over more abstract issues in pure science this has often been the case as new data become available. One such example is the controversy over the continental drift theory that lasted for almost half a century but remained an issue mainly for the earth scientists, without immediate political, economic, or social aspects and thus was largely confined to the scientific domain.[14] When such aspects are present, and where the scientific dispute is embedded in a technological, environmental, or ideological context, solving tends to take other forms, including the intervention of nonscientific actors and stakeholders, as in the case of nuclear energy.[15]

---

[13] Ernan McMullin, "Scientific Controversy and Its Termination," in *Scientific Controversies: Case Studies in the Resolution and Closure of Disputes in Science and Technology*, ed. Hugo Tristram Engelhardt and Arthur Leonard Caplan (Cambridge: Cambridge University Press, 1987), p. 51.

[14] Henry Frankel, "The Continental Drift Debate," in *Scientific Controversies: Case Studies in the Resolution and Closure of Disputes in Science and Technology*, ed. Hugo Tristram Engelhardt and Arthur Leonard Caplan (Cambridge: Cambridge University Press, 1987), p. 203.

[15] On the nuclear energy controversy, see Spencer R. Weart, "Nuclear Fear: A History and an Experiment," in *Scientific Controversies: Case Studies in the Resolution and Closure of Disputes in Science and Technology*, ed. Hugo Tristram Engelhardt and Arthur Leonard Caplan (Cambridge: Cambridge

Ernan McMullin identifies three main categories of how scientific disputes end: resolution, closure, and abandonment:[16]

- *Resolution* means that the controversy is resolved in the sense that both sides accept one of the contesting views, or a modified middle view. This way to terminate a controversy is in line with the self-image of science as open, critical, and free from nonscientific prejudice.
- *Closure* means that the controversy is terminated through the intervention of an external force, one not necessarily bound by the actual merits of the case. This closure can take the form of a court ruling, the withdrawal of funds, or a reorganization of research institutions. Closing the controversy does not solve it, and the disagreement could still exist under the surface.
- *Abandonment* appears in situations where the contested issue disappears, either because the contenders lose interest, or grow old and die, or because the issue becomes less relevant and is bypassed by other discoveries or theories.

One of the prolonged and intense scientific disputes, displaying both the devastating effects and the insufficiency of closure as a method to terminate a fundamental scientific controversy is the rise and fall of T. D. Lysenko and the subsequent destruction and eventual rebuilding of Soviet biological research from the 1930s to the 1960s.[17] The agronomist Lysenko and his followers managed to establish hegemony for the biological quasi-science of "lysenkoism," based on the theory that plants and livestock could be improved through "re-education" and that acquired characters could be inherited. The case was supported by research results presented by Lysenko and his followers, rhetorical links to Marxist-Leninist dialectics, and not least the promise of deliveries in the form of grossly improved productivity in Soviet agriculture. The hegemony was achieved in a fierce and literally deadly scientific controversy, in which Lysenko from the late 1930s onward denounced his opponents as representatives of a "bourgeois" science, and as such, enemies of the people, an accusation making them fair game for a witch-hunt by the NKVD (Peoples Commissariat for Internal Affairs), the Soviet internal security service at the time. The flawed theories of Lysenkoism, and the dead hand it lay over Soviet biological research had a severe impact on Soviet agriculture over several decades.

University Press, 1987), and M. Bauer, *Resistance to New Technology: Nuclear Power, Information Technology and Biotechnology* (Cambridge: Cambridge University Press, 1997).

[16] McMullin, "Scientific Controversy and Its Termination."

[17] Medvedev and Lysenko, *The Rise and Fall of T. D. Lysenko*, and David Joravsky, *The Lysenko Affair* (Cambridge, MA: Harvard University Press, 1970).

While the Lysenko affair in many respects represents an extreme and almost bizarre form of scientific controversy, reflecting the conditions of science and arts under extreme totalitarianism and terror, it still highlights some elements of more general validity. One of these is the relation to power. Lysenkoism, or any similar form of non-science based on metaphysic beliefs, could never have survived, let alone triumphed over genetics, had it not been for the ability of its prophet to win the ear and support of the political power. Lysenko framed himself as the underdog practitioner who knew the *real* agricultural problems better than fancy "bourgeois," hence counter-revolutionary, scientists.[18] Lysenko also, through this link to power and the hegemony of the "general line" of the Communist Party, could use the state-controlled media both to trumpet his own version and to criticize his opponents as his power increased, preventing their ideas from being published at all. This pattern of directly or unintentionally mobilizing political or other external support in a scientific dispute to intimidate and silence opponents is hardly specific to the Soviet case or for the scientific domain, as discussed below.

Controversies in Intelligence

Controversies in intelligence appear in a setting that is different from science in both an epistemological and organizational sense. The scientific culture and the intelligence culture, while sharing some characteristics, nevertheless, as discussed in the second chapter, are distinctively different in a number of ways. Both strive, on a formal and philosophical level, to uncover, explain, and assess elements of what is assumed to be observable reality. Both have a common, though perhaps not universally accepted, ethos of impartiality and honesty in terms of facts and findings. Both detest fraud as unprofessional and destructive for the credibility of all knowledge-producing activities. Intelligence, however, keeps its fingers crossed behind its back when it comes to fraud intended as disinformation and deception. Intelligence, furthermore, uses means and produces outputs that would violate research ethics, while some such practices instead are perfectly in line with a different intelligence ethic. The duty of intelligence is to protect competitive advantages in terms of technology, software, or human resources, in the process often producing results that to lesser or higher degree are sensitive or damaging for individuals, states, or the international community.

The main difference is perhaps not the issues as such but rather the ways these issues are dealt with, the functions of the intelligence machinery, its legal

---

[18] Joravsky, *The Lysenko Affair*, pp. 187–190.

basis, and not least its cultural setting—the strange and somewhat elusive intelligence culture described by Michael Herman as a combination of a sense of being different, of having an exclusive mission, and the multiplying factor of secrecy.[19] Many researchers also share a sense of being different (sometimes even superior) as well as a sense of mission, not only in terms of scientific progress per se but also toward society and issues on a global scale, some of which the researchers perceive as both more over-arching and more important than narrowly defined intelligence requirements. Secrecy and its consequences in terms of closeness, compartmentalization, the need-to-know principle, protected sources, and the operational link to decision making thus stand out as the main differences, producing a very different setting for the generation and handling of intelligence controversies, compared to those within science.

A first observation is that in intelligence, being a closed knowledge-producing entity, many or possibly most controversies that do emerge remain out of sight of the public domain and are in many cases contained within organizations and thereby invisible not only while they still exist but also in retrospect. They might leave only traces in the archives—for example, the increasing controversy in the US intelligence community over the NIEs of the future developments in the Soviet Union from the late 1980s onward. But they might also stay confined to arguments within intelligence services and remain undocumented in intelligence assessments, only remembered by those drawn into the dispute. One example is an undercurrent of critique in Swedish military intelligence in the second half of the 1970s directed toward the dominating concepts of détente and of the Soviet Union having renounced any aggressive ambitions toward the West. As a competing hypothesis, this critique was neither strange nor unique. It merely mirrored a growing and increasingly loud concern among traditionalists that things were not quite what they seemed to be after the 1975 Helsinki Treaty, and that the West, while drinking innumerable toasts to peace and cooperation, was slowly but steadily marching into a trap. This dissenting view, in contrast to the case in the United States, never appeared in intelligence production; it remained a critical, undocumented undercurrent confined to casual corridor meetings, coffee room discussions, and confessions behind closed doors. This unresolved and almost invisible controversy played an important role in the re-definition of Sweden's geostrategic position in the early 1980s in the second Cold War and under the influence of the so-called submarine crisis with the Soviet Union.[20]

---

[19] Herman, *Intelligence Power in Peace and War*, pp. 327–329.

[20] For the Submarine Crisis, see Fredrik Bynander, *The Rise and Fall of the Submarine Threat: Threat Politics and Submarine Intrusions in Sweden 1980–2002* (Uppsala: Uppsala Acta Universitatis Upsaliensis, 2003).

Against the background of the differences between science and intelligence, it is not surprising that controversies in intelligence in many cases are avoided or simply repressed; they are hardly looked on with favor in a production-oriented and (supposedly) fact-finding activity. Controversies would, as in a bureaucracy or a military organization (of which intelligence often is either or both), be seen as an indication of weak leadership or insufficient external guidance and supervision. This is mirrored in the often-quoted line by the chief of Israeli Military Intelligence, Major General Eli Zeira, before the 1973 Yom Kippur war that he wanted to give the decision makers a clear, straightforward assessment and that intelligence should be either unambiguously right or unambiguously wrong. Unfortunately, in the ultimate test against reality, he turned out to be right only in that his faulty assessments were astonishingly unambiguous.[21]

We can call this outcome prima facie closure; once the sky has fallen down, the debate over the assessments of the risk for the sky falling down is over, and the time for excuses, finger-pointing, commissions, and reorganizations begins. We discuss several of the intelligence failures in this book and the character of prima facie closure, with the subsequent consequence that the underlying explicit—or implicit—controversy was not automatically settled. This becomes visible in the instances where intelligence failures include other elements, or where the alleged failures are not clear-cut failures or not even failures at all. Richard K. Bett's postmortem of the US 2002 NIE on Iraqi weapons of mass destruction illustrates the pitfall of the prima facie closure and is further discussed in Chapter 8.[22]

If intelligence controversies cannot be avoided or preempted, they are often terminated by closure—that is, by intervention down the command chain or through transforming a dispute over substance into one of formalities and regulations. The attempt by the head of the Jordanian desk in the Israeli intelligence service to challenge the assessment made by the Egyptian and Syrian desks in the final stages before the 1973 war was viciously repressed by their superior, the head of the research department, on the grounds that the officer at the Jordanian desk not only had interfered in something that was none of his business but also had done so based on incoming intelligence that he was not cleared to see according to the need-to-know principle. The whole issue was thus transformed into a matter of discipline and breach of security regulations, and the eventual factual merits of his case were never tested.[23] A similar

---

[21] On Zeira's and the Research Department's assessment prior to the war, see Bar-Joseph, *The Watchman Fell Asleep.*

[22] Richard K. Betts, *Enemies of Intelligence: Knowledge and Power in American National Security* (New York: Columbia University Press, 2007).

[23] Bar-Joseph, *The Watchman Fell Asleep*, pp. 90–92.

formalistic anti-intellectual attitude is not entirely uncommon in academia either, especially in domains and institutions preoccupied with the process of establishing a field as "normal science," or with well-developed mechanisms for preserving an existing normal science.

Abandonment as a way to terminate a scientific dispute, finally, is possibly more common in intelligence than in science. Intelligence does collect, and process, a vast mass of information, but it is with some notable exceptions not a process of cumulative knowledge production (see Chapter 7). Most of the collected material is relevant for only a limited period of time, and once the focus shifts it becomes "history." Old assessments are not regarded as useful and relevant, even in instances when actually they are, and too often new data with the same conclusions are developed from the ground up.[24] Controversies can be important to clarify as a part of an after-action review if lessons with a wider validity are to be learned, but aside from that, what is bygone is bygone. Faulty estimates of Soviet force postures from the Cold War are primarily a matter for intelligence historians and authors of textbooks.

Science versus Metaphysics—Or Who's the Expert?

Several of the more famous and intense scientific controversies have emerged over the difference between what the scientific communities and their institutions regarded as science and what is (or should be) discarded as nonscientific beliefs or misconceptions. The intense debate on global warming is just the latest of a number of often very intense and emotional clashes over not only interpretations of observations and the logic of inferences, but the legitimacy of one's claims to make such interpretations and inferences in the name of science. Some of these controversies erupt between elites and counter-elites, between the established scientific community and competing entities, whether in a political structure as was the case with the Communist Party Central Committee in the Lysenko affair, or from a popular movement perceiving science as bought or perverted by political influence or commercial interests.[25] But the bulk of the

---

[24] Robert Jervis gives a small example of this in his account of the CIA's assessments on Iran prior to the fall of the Shah in 1979. Older assessments from 1964, that Jervis suspected could be of relevance, had been removed from the archive and sent to "dead storage" from where it would have taken weeks to get them retrieved. Jervis, *Why Intelligence Fails*, p. 25.

[25] One example of this is the widespread popular disbelief in official assessments of the risks from the fallout from the Chernobyl nuclear disaster in 1985. See Angela Liberatore, *The Management of Uncertainty: Learning from Chernobyl* (Amsterdam: Gordon and Breach, 1999).

controversies can nevertheless be found within the scientific community, the universities, and the disciplines themselves. These sometimes are intellectualized territorial or personal conflicts over the control and boundaries of a discipline or research field, access to funding, or who is awarded individual chairs in a university department. Nothing of this should be surprising; the academy is just another arena for well-known (and well-studied) patterns of competitive social behavior. One of the most powerful arguments in this competitive setting is probably to accuse opponents not only of being inferior scientists in terms of quality, citations, and ability to attract funding but also as representing a dilution of science, the use of unscientific methods, and the presentation of ill-founded results. Going one step further, those criticized can be branded as representatives of non-science and duly excommunicated.

But where should the line be drawn? Popper focuses on the degree of testability and differentiates between well-testable theories, hardly testable theories, and non-testable theories. Of these, the former two belong to the realm of science while the third belongs to the realm of metaphysics.[26] There is, in Popper's words, a line of demarcation between the two realms. However, this line of demarcation cannot, according to Popper, be drawn too sharply, since much of modern science has in fact developed from myths and misconceptions: "It would hardly contribute to clarity if we were to say that these theories are non-sensical gibberish in one stage of their development, and then suddenly become good sense in another."[27]

Furthermore, some statements are testable and belong to science, while their negations are not, and thus must be "placed below the line of demarcation" as constituting metaphysics. Popper illustrates with the statement that perpetual motion exists, which is testable, while the negation, that there are no perpetual motion machines, belongs to metaphysics. Yet Popper argues that "isolated purely existential statements" should not be discarded as falling outside the scientist's range of interest.[28] Popper's observation about the history of science is that it is a gradual cognitive process from beliefs and preconceptions toward observation, increased testability, and the development of theories based on the principle of refutability. We cannot, according to his arguments, know something with any certainty unless we know *why* we know it. But given the cognitive process described, there might nevertheless be something out there.

The problem surrounding Popper's "line of demarcation" between science and metaphysics, and the ambiguous role of the latter is illustrated not only in scientific development but perhaps even more visibly in the domain of intelligence,

---

[26] Popper, *Conjectures and Refutations*, p. 257.
[27] Ibid.
[28] Ibid., pp. 257–258.

where the criteria of refutation often are difficult and sometimes impossible to fulfill, at least in the time-frames in which intelligence normally operates. One of the most discussed cases of 20th-century intelligence analysis—the ill-fated US assessments of Soviet intentions on Cuba in 1962—illustrate this; the case has been mentioned in Chapter 3.[29]

In August 1962, the newly appointed director of the Central Intelligence Agency, John A. McCone, raised the issue that the Soviets might try to place nuclear missiles on Cuba to offset the US superiority in terms of strategic weapons, and he saw the deployment of Soviet air defense systems on the island as verification. Being an amateur in intelligence matters, McCone's hunch was regarded as belonging below the line of demarcation, a not very uncommon eventuality in the interaction between specialists and non-specialists, whether in intelligence or in science. However, given McCone's position, his hunch could not just be discarded out of hand, and the result was reflected in the Special National Intelligence Estimate (SNIE)—in many respects a revealing exercise in which a hypothesis from below the demarcation line was duly dealt with by the experts convinced they were operating above the line. Based on observations in terms of previous Soviet behavior and common sense logic, what the Soviets ought to regard as being in their own interest, the hunch was put back where it belonged, only to reemerge a few weeks later, first as a disturbing and illogical anomaly in the intelligence flow and then confirmed beyond doubt as a prima facie termination of an intelligence controversy.[30]

The problem in the Cuban case, as observed by many subsequent commentators, was that the analysts mistook their *provisional* knowledge for *actual* knowledge, and filled in the blank spots with assumptions of how things should be and how the Soviets "must" think. While appearing, and certainly perceived, as solid ground, these statements were in fact metaphysical and just as much below the line as McCone's ideas, although as it turned out, in a more dangerous mode as it was ignorance perceived as verified knowledge. The Cuban case also illustrates the practical implication of Popper's point about perpetual motion. The hypothesis that nuclear missiles were deployed or were in the process of being deployed was in principle testable, although from an intelligence collection perspective the task would be difficult and time-consuming if not sharply different from a vast number of other intelligence tasks regarding development,

---

[29] SNIE 85-3-62, *The Military Buildup in Cuba*, September 19, 1962, in Mary S. McAuliffe, *CIA Documents on the Cuban Missile Crisis, 1962* (Washington, DC: Central Intelligence Agency, 1992). For the background of the SNIE, see James G. Blight and David A. Welch, *Intelligence and the Cuban Missile Crisis* (London: Frank Cass, 1998).

[30] On the employment of common sense and pattern analysis in the Cuban case, see Agrell, *Essence of Assessment: Methods and Problems of Intelligence Analysis* (Stockholm: National Defence College, Center for Asymmetric Threat Studies, 2012).

deployment, and proliferation of weapons systems. The *negation,* however, was not testable, given the limited means of intelligence collection available.

This phenomenon reappeared on a wider scale prior to and after the 2003 Iraq war, where the hypothesis that Iraq possessed WMD was, from the background of previous observations and common sense, perceived as testable, with vast and as it turned out futile attempts to find (or create) verification. The negation, the testing of the hypothesis that Iraq did *not* possess any WMD, was not really attempted and would not have been achievable under the prevailing circumstances: on that score, recall the line by then-Secretary of Defense Donald Rumsfeld that the absence of evidence was not evidence of absence. It took a year of post-defeat excavation to reach the relatively reliable assessment that the fact nothing had been found meant that there was nothing to find.[31]

Turning back to Popper, we can see that his "line of demarcation," while philosophically clarifying, nevertheless is of limited use in a context where knowledge production is conducted under considerable uncertainty and policy interaction. Sociologist Thomas Gieryn suggests the concept "boundary work" to describe the way the scientific domain is defined in the latter, more complex context. Science, he writes, is not static: "its boundaries are drawn and redrawn in flexible, historically changing and sometimes ambiguous ways. This process is conducted by scientists in order to contrast it favourably to non-science intellectual or technical activities."[32] "Boundary work" as a concept is increasingly relevant in the fields where science becomes engaged in the exploration and explanation of complex problems with a high degree of policy relevance, and where the complexity not only refers to the matter of substance, but also the policy recommendation, and their potential impact on the object of study, the "blue" dimension further discussed in Chapter 6.

Scientific Warning and Policy Response: Acid Rain and Forest Death

Rachel Carson's book *Silent Spring* in 1962 was one of the first instances of a scientific warning of environmental threats, attracting global attention and thereby affecting policy.[33] The role of the scientists was obvious; they had the

---

[31] On the finding of the Iraqi Study Group and their implications for the evaluation of previous intelligence assessments, see Betts, *Enemies of Intelligence.*

[32] Thomas F. Gieryn, "Boundary-Work and the Demarcation of Science from Non–Science: Strains and Interests in Professional Ideologies of Scientists," *American Sociological Review* (1983): 781–795.

[33] Rachel Carson, *Silent Spring* (Boston: Houghton Mifflin, 1962).

ability to collect data, the knowledge to interpret them, and not least the undisputable authority to underpin their conclusions as well as implicit or explicit policy recommendations. But *Silent Spring* had one further aspect: that of allowing readers to visualize an impending disaster. "Boundary work" in this and other existential issues, like the arms race and nuclear energy, was about the duty of the scientists to reach out beyond the mere academic limits.

In the 1980s, acid rain became an important environmental issue both in North America and in Western Europe.[34] The issue was raised and named by scientists in the 1970s, and the scientists were instrumental in framing it for the general public and the policymakers. In the United States this resulted in a national research program designed not only to monitor and assess acid precipitation but also to assist lawmakers with scientific expertise in hearings.[35] On the political level, the issue of acid rain and subsequent limitations of emissions from industry was controversial and deeply polarized. The scientists encountered not a single audience but two distinct sides in a policy process that was highly adversarial. It was not just a question of speaking truth to power, but what truth and to which power? Scientific objectivity was put under strain, both in terms of expectations from policymakers and also the ambition of scientists to—at least momentarily—leave the traditional scientific role and give policy advice.[36]

The scientist's role became problematic not only due to the pressure from expectations and the form of interaction between science and policy in hearings; the complexity of the issue, and uncertainty regarding the conclusions on causes and effects, blurred the concept of knowledge. If the scientists themselves did not know for sure, then just what was the basis for their claim of knowledge? And who was then the expert, after all? Another, and as it turned out increasingly important, factor was that of urgency. Acid rain could have serious environmental impacts, but once limited, these effects were assumed to gradually decrease. In Central and Northern Europe, however, the acid rain

[34] Due to early discoveries of Svante Odén, professor in Soil Science, acidification was acknowledged as a major environmental threat in Sweden in the late 1960s and emissions were drastically reduced through legislation. However, attempts to raise the issue internationally proved unsuccessful, and Swedish and Norwegian efforts came to nothing until the problem reached the policy agendas in the United States and Western Germany. See Lars J. Lundgren, *Acid Rain on the Agenda: A Picture of a Chain of Events in Sweden, 1966–1968* (Lund: Lund University Press, 1998).

[35] Stephen Zehr, "Comparative Boundary Work: US Acid Rain and Global Climate Change Policy Deliberations," *Science and Public Policy* 32, no. 6 (2005): 448.

[36] Zehr (2005) and Lundgren (1998) point at the risk of scientists, in dealing with acid rain, in becoming too much guided by what should be done, and gravitating toward what could be done, with the associated risk of simplification and the resulting tendency toward monocausal thinking. Ibid., pp. 448–450, and Lundgren, *Acid Rain on the Agenda*, p. 292.

issue became framed in a different way beginning in the early 1980s. It was connected to an observed, widespread decline in forest vitality, especially in Central Europe (Czechoslovakia, East Germany, Poland, and West Germany). In West Germany alone, a survey conducted in 1983 revealed—or seemed to reveal—that 34 percent of all forested land was affected.[37] Due to the long life span of trees, observations could indicate disastrous effects of exposure that had accumulated over a long time.

The discovery of forest death had a long pre-history, with known intervals of decline in growth over the centuries. The link with manmade influences, especially air pollution over remote distances, was not highlighted until the 1970s, when observations from several locations in Europe were connected and a common cause was sought. The main breakthrough came when the German forest ecologist Bernhard Ulrich put forward the hypothesis of a combined cumulative effect of increased acidity in precipitation, leading to altered soil chemistry causing the death of the tree's fine roots.[38] This hypothesis connected the acid rain debate and a seemingly new ecological (and potentially economic) threat. The forest death concept was rapidly established in the German public debate under the designation *Waldsterben* and came to activate earlier misgivings about the environmental impact of long-range air pollution in other countries as well. Policy response was swift: in the early 1980s the German government reversed its previous stance resisting international reduction of sulfur emissions, and instead actively promoted an multi-national agreement.[39]

Not all scientists shared the alarmist perception of imminent catastrophic forest damage; that was particularly so for those who had studied forest disease and growth over time. Empirical forestry research, however, was by definition a slow process, and publication in scientific journals suffered from a considerable time lag. And once the forest death had been established in media, there was little interest in critical remarks or data casting increasing uncertainty.[40] The rapid, and almost total, impact of the initial scientific warnings on media, public opinion, and policy placed the scientific community in a difficult situation in which there was a demand for rapid scientific results to initiate countermeasures. In the Swedish debate, the concept of forest death had an impact similar to that in Germany even though data on growth and damage was far

[37] Jan Remröd, *Forest in Danger* (Djursholm: Swedish Forestry Association, 1985).

[38] B. Ulrich, "Die Wälder in Mitteneuropa. Messergegnisse Ihrer Umweltbelastung, Theorie Ihre Gefärdung, Prognosen Ihre Entwicklung," *Allgemeine Forstzeitschift* 35, no. 44 (1980): 1198–1202.

[39] Nils Roll-Hansen, *Ideological Obstacles to Scientific Advice in Politics: The Case of "Forest Death" from "Acid Rain"* (Oslo: Makt–og demokratiutredningen 1998–2003, 2002), p. 8.

[40] Ibid., p. 9.

less alarming. Research programs were designed to study the complex ecological interaction behind forest death, not to question the concept or cast doubt on the mathematical models for calculating future decline. Scientists were also subjected to a demand pull for results that could be used by the government in international negotiations, and the repelling effect from a polarized public debate where critics of the acid rain–forest death interpretation were seen as acting on behalf of industry and against an alternative "green" policy.[41]

In the end, the refutation of the acid rain–forest death hypothesis turned out to be a gradual and slow process. Several large-scale monitoring projects were initiated in the early 1980s, and after a few years they started to produce empirical results, published in reports and scientific journals. An assessment of the state of the art in the mid-1980s, published by the Nordic Council of Ministers, concluded that the so-called cumulative stress hypothesis could not be proven and that there were several possible explanations for the decline observed in the Central European forest. Acid rain was not ruled out as a culprit, but the main recommendation was more research on the dynamics of the forest ecosystem and the establishment of more accurate survey methods to avoid subjective assessments of decline.[42] Within the European Communities a forest damage inventory system was set up in 1987, and annual reports were published. The results did not fit the alarming assessments made in the first half of the 1980s, based on data from more scattered locations. During the first three years of the inventory, no clear change in vitality was found for the majority of species, and no increasing percentage of damaged trees could be recorded. Furthermore, due to insufficient data, it was not possible to establish a relationship between air pollution and forest damage in the European Community. At some more limited sites, where data were available, no such relationships were found.[43]

A first policy shift came as early as 1985, when the German government decided to downgrade the threat and changed the vocabulary from forest death (*Waldsterben*) to forest damage (*Waldschaden*).[44] With growing uncertainty

[41] Anna Tunlid, "Ett Konfliktfyllt Fält: Förtroende Och Trovärdighet Inom Miljöforskningen (A Field of Controversy: Trust and Credibility in Environmental Research)," in *Forskningens Gråzoner – Tillrättaläggande, Anpassning Och Marknadsföring I Kunskapsproduktion* (The Grey Zones of Research – Adjustment, Adaption and Marketing in Knowledge Production), ed. Wilhelm Agrell (Stockholm: Carlssons, 2007).

[42] Lars Moseholm, Bent Andersen, and Ib Johnsen, *Acid Deposition and Novel Forest Decline in Central and Northern Europe: Assessment of Available Information and Appraisal of the Scandinavian Situation: Final Report, December 1986* (Copenhagen: Nordic Council of Ministers, 1988).

[43] Commission of the European Communities, Directorate–General for Agriculture, *European Community Forest Health Report 1989: Executive Report* (Luxembourg: Office for Official Publications of the European Communities, 1990), pp. 23–24.

[44] Roll-Hansen, *Ideological Obstacles to Scientific Advice in Politics*, p. 18.

about both the extent and duration of the damage, as well as about the causal link to acid rain, the issue gradually lost public and policy attention, though in some countries like Switzerland and Sweden the perception was more long-lived and upheld by persistent scientific consensus. In the Swedish case, large-scale countermeasures had been initiated to prevent the development of widespread damage to the forests. The acid rain–forest death hypothesis finally collapsed 1992 when comprehensive surveys failed to reveal a continuous epidemic and instead could be seen as signaling an increased growth rate. In a broader perspective, the alarming observations from the late 1970s appeared as part of a normal fluctuation due to factors like weather and insect pests.[45]

In retrospect, the acid rain–forest death alarm in Western Europe stands out as a major scientific failure, as well as a media-driven over-reaction. It is hard to regard the rise and fall of the forest death concept as something other than a flawed scientific warning. True enough, the warning did work, or worked too well; the trouble was that it turned out to be unfounded, a situation similar to cases of over-warning in intelligence, associated with a subsequent cry-wolf syndrome.[46] In one sense, though, the self-correcting mechanisms of the scientific community did work; once more reliable and representative data were available, the hypothesis was undermined and finally collapsed. Nils Roll Hansen compares this with the Lysenko case where this process took three decades, while in the forest death case it took only one.[47] The self-correction mechanisms of science were not eliminated as they had been under Stalin but were nevertheless slow and retrospective, to some extent simply due to the methods of data collection over time. Moreover, there were other distorting mechanisms at work. One was scientific consensus and research funding. As forest death exploded on the public agenda, it became more difficult for dissenting voices to be heard, and with large-scale research funding motivated by the need to counter the threat, dissent was definitely not rewarded—or rewarding. Scientific criticism was dampened and a kind of bubble effect emerged, in which scientists could feel that it was almost their duty to prove the existence of forest death.[48]

---

[45] Ibid., pp. 4–5.

[46] One of the well-known cases in intelligence literature is the successive US warnings for war with Japan, starting in summer 1940 and by December 1941 resulting in a warning fatigue. See Wohlstetter, *Pearl Harbor: Warning and Decision*. The slow Israeli response to the military buildup prior to the October 1973 war is often explained with reference to the over-reaction in connection with a crisis in Lebanon in May–June, resulting in costly and, as it turned out, unnecessary mobilizations.

[47] Roll-Hansen, *Ideological Obstacles to Scientific Advice in Politics*, p. 36.

[48] Ibid., p. 22, quoting a study by the Swiss researcher Wolfgang Zierhofen on the forest death issue in Switzerland.

The rise and collapse of the forest death hypothesis in several respects resembles a classical intelligence failure. The combination of uncertainty and potentially alarming consequences compelled the scientific community to gear up and to communicate preliminary observations at an early stage. Once this was done, media logic and policy momentum took over. The scientists had to deliver and continue to deliver a process that raised the revision threshold as the price of being wrong became too high. Uncertainty and consensus proved to be an unfortunate combination, as well as the suspicion of being biased by ideology or non-scientific partisan interests. Forest death never reached the level of a paradigm, and the "normal science" conducted in the crash research programs was predestined to sooner or later undermine the basic hypothesis. Nevertheless, there were prolonged instances of paradigmatic defense and rearguard actions by key researchers. One Swedish biologist maintained that it was because of the warnings and subsequent countermeasures that the environmental impact had turned out as favorable as it did, thereby probably unknowingly walking in the footsteps of Sherman Kent dismissing the flawed Cuban estimate on the grounds that events had proved that the CIA was right and it was Khrushchev who had been in error! The Iraqi WMD case did not allow for the same kind of last-ditch paradigmatic stand.

Global Warming and the Impact of Consensus under Uncertainty

Both the acid rain and the subsequent forest death issues could be seen as predecessors to the rise of global warming. Several of the scientists who started their environmental engagement on the acid rain issue proceeded to work with global warming.[49] Just as with acid rain and forest death, global warming was discovered by the scientists, but the process was more drawn out, with initially contradictory interpretations and a gradual process toward an emerging consensus, influenced by global warming becoming a major international policy issue in the late 1980s. The creation of the Intergovernmental Panel on Climate Change (IPCC) in 1988 was the major benchmark in this process.

The early scientific controversy over climate change was dominated by widespread uncertainty, not only about data on contemporary climate trends but also about the glaciological pre-history, and how the climate functioned and fluctuated as an immense complex system on a planetary scale. By the mid-1960s, existing

---

[49] Richard S. J. Tol, "Regulating Knowledge Monopolies: The Case of the IPCC," *Climatic Change* 108 (2011): 828.

theories were challenged by the hypothesis that climate change might not come about through a gradual process over thousands of years, but rather through sudden shifts over hundreds of years.[50] One of the issues raised at an early stage was the potentially catastrophic impact on the ocean level of a melting, or rather disintegration, of the so-called West Antarctic ice shelf.

However, in the 1970s the scientists were still divided as to the direction of a sudden climate change: would it be a shift toward global warming or a new ice age? At this stage the issue was still confined to an internal scientific dispute; it concerned highly theoretical matters, and any potential impact was far into the future. Climate change was in this sense invisible, and as such a public non-issue. The extraordinary novelty in this case, observes Spencer R. Weart, was that such a thing became a political question at all.[51] This process started in the mid-1980s. Assisted by a heat wave in the United States in the summer of 1988, scientists and politicians managed to get maximum public attention to a congressional hearing, where the leading scientist James Hansen stated "with 99 percent confidence" that there was a long-term warming trend under way, and that he strongly suspected the increased emission of carbon dioxide—the greenhouse effect—to be behind this warming.[52]

This transfer from the scientific to the public domain, and the politicization in the original sense of the word (see Chapter 8), was due to several interacting factors. A fundamental precondition was that the time was ripe, with debates over nuclear winter and acid rain having paved the way.[53] And with the nuclear arms race winding down, there was room for other global concerns. Scientific and policy entrepreneurs also played a critical role in the process by translating incomprehensible scientific data into visible threat perceptions and media-friendly soundbites.

Climate research had from the very onset been multi- or rather trans-disciplinary. The complex mechanisms could be analyzed neither solely within any existing discipline nor through the combination of methods and theories from several existing disciplines. A new set of scientific tools had to be created more or less from scratch, and this, the development of climate modeling, became a central concern for IPCC, and the basis for their successive assessments of warming, its implications, and subsequent

---

[50] For the epistemology of the global warming theory, see Spencer R. Weart, *The Discovery of Global Warming* (Cambridge, MA: Harvard University Press, 2003).

[51] Ibid., p. 153.

[52] Ibid., pp. 154–157.

[53] A similar link was observed between forest death and acid rain, where the former functioned as an eye-opener on the actual impact of acidification on the environment. Forest death was thus "helpful" for the acidification question, providing a strong argument for further measures. Lundgren, *Acid Rain on the Agenda*, p. 289.

policy recommendations.[54] The IPCC itself was not a research institute but rather has been described as a hybrid organization[55] overlapping science and policy, with both scientists and governmental representatives. It has no scientific staff and instead relies on a vast network of volunteer researchers, thereby operating in a fashion similar to an nongovernmental organization or NGO. IPCC's work followed an intelligence-like production cycle, with Assessment Reports as the major undertaking, published in 1990, 1995, 2001, 2007, and 2014.[56]

The Intergovernmental Panel on Climate Change was created jointly by the United Nations Environmental Program (UNEP) and the World Meteorological Organisation (WMO), in many ways incorporating the multi-lateral negotiation culture of the United Nations (UN) system and other international organizations. IPCC was different mainly in the dominant role played by the scientist network and the aim of producing assessment in a joint process with scientists and government representatives engaged in the review and approval process.[57] Any disputes on interpretations within the system are therefore set to be solved mainly by closure through negotiations and consensus. Failure to reach consensus would, as in the UN system, lead to a non-result, in this case, the deletion of an assessment.[58]

The creation of the IPCC resulted in a gradual termination of the wider scientific controversy by closure. The Intergovernmental Panel was dominated by scientists who not only came to agree on global warming as an increasingly verified hypothesis but also on the causal link to human activities, and on a common sense of duty to explain and communicate this to the policymakers

---

[54] Intergovernmental Panel on Climate Change, *Principles Governing IPCC Work* (Vienna: IPCC, 1998).

[55] Zehr, "Comparative Boundary Work," pp. 454–455.

[56] A preliminary report was published in September 2013 in Intergovernmental Panel on Climate Change, *Climate Change 2013—the Physical Science Basis: Working Group I Contribution to the Fifth Assessment Report of the IPCC (Preliminary Report)* (Cambridge: Cambridge University Press, 2013).

[57] There is a vast literature on the IPCC by participants and social scientists. Aspects of "boundary work" between science and policy is dealt with by Zehr (2005), a critical account of the dominating role of IPCC is given in Tol (2011), while a review of the research is provided in Mike Hulme and Martin Mahony, "Climate Change: What Do We Know about the IPCC?" *Progress in Physical Geography* 34, no. 5 (2010): 705–718.

[58] One example of this is the growing uncertainty over estimates on the stability of the western Antarctic ice sheet prior to the drafting of Assessment Report 2007 (AR4). The inability to reach consensus on how to interpret new data for short- and long-term predictions of the impact on sea level rise resulted in no prediction at all being delivered. See Jessica O'Reilly, Naomi Oreskes, and Michael Oppenheimer, "The Rapid Disintegration of Projections: The West Antarctic Ice Sheet and the Intergovernmental Panel on Climate Change," *Social Studies of Science* 42, no. 5 (2012): 709–731.

and a wider audience.[59] Climate critics or skeptics were increasingly marginal-ized, often with reference to their partisan interests, and their dissenting views were not published in the leading scientific journals.[60] It is significant that the chairman of the IPCC, in commenting on suggested improvements in the panel procedures to include dissenting views, felt that he had to underline that he did not use the term "critics" in a derogatory way.

The overall goal of the IPCC, as outlined in the governing principles, was to assess "on a comprehensive, objective and open basis" information relevant to the understanding of the risk of human-induced climate change.[61] As stated in the regulation of IPCC procedures, the various bodies should "use best endeav-ors to reach consensus."[62] The focus on consensus reflected both the instrumen-tal role of the output and the hybrid structure, in which knowledge is negotiated through an elaborate process involving not only scientists but also governmen-tal representatives.[63] Falsification efforts are just as unhelpful here as in many instances in intelligence. In this respect the epistemological foundation of the IPCC was clearly one of positivism; the climate issue is perhaps complex, but it is knowable and the road forward thus depends on the scale and duration of the efforts as well as the rigor with which procedures were adhered to. Seeking consensus was, from this standpoint, not a problematic goal; on the contrary, the more data available and the more research efforts merged, the more likely it would be that the consensus assessments would approach the goal of correctly describing and modeling global climate dynamics. The often-made reference to the consensus of 2,500 of the world's leading scientists underpinned the validity of the constructed "truth."[64] The epistemology of the IPCC approach resembles the one that dominated intelligence doctrine during the Cold War and after 911, with a heavy focus on collection surge and multiple layers of bureaucratic overlay, based on the presumption that future threats are knowable and that intelligence organizations should strive toward this goal.

Especially after the release of the Forth Assessment Report 2007 and a subsequent debate over errors and biased editing, the IPCC peer review sys-tem, overall guidance, and dominating role in climate research came under debate, focusing on the downside of consensus building and the handling of uncertainty. As an answer to the critique, an independent review of the IPCC processes and procedures was conducted by the InterAcademy Council in 2010. While not constituting a postmortem, the review came close to being

[59] Hulme and Mahony, "Climate Change," p. 711.
[60] Weart, *The Discovery of Global Warming*, pp. 166–167.
[61] Intergovernmental Panel on Climate Change, "Principles Governing IPCC Work."
[62] Ibid.
[63] Hulme and Mahony, "Climate Change," pp. 710–712.
[64] For the consensus building, see Ibid.

a scientific equivalent to an intelligence oversight evaluation. Dealing mainly with procedural matters, the review nevertheless warned IPCC against stagnation and recommended stronger executive and editorial leadership. But the review also underlined the importance of ensuring that "genuine controversies" were adequately reflected, and that all Working Groups should describe the level of uncertainty in a uniform fashion, which had not been the case.[65]

The critique of the methods and procedures employed by the IPCC reflects to some extent the impact of the "little divide" between hard sciences on the one hand and social sciences and humanities on the other. The dominating role of the former resulted in a marginalization of the latter and a dominance of quantitative methods over qualitative. That, in turn, led to the subsequent weakness in the ability to communicate non-quantifiable uncertainties that the InterAcademy review identified. A presumably scientific effort here encountered one of the chronic problems of drafting and communicating intelligence assessments (see Chapter 3).

As a result of the critique of the assessments presented by the IPCC in the fourth report, as well as the increasing diversity of the research literature, the chapter teams preparing the Fifth Assessment Report developed what was described as a more sophisticated model to treat uncertainty. Perhaps the most interesting aspect of this process is the way the teams dealt with the divide between natural and social sciences, where the former by tradition are more familiar with quantitative descriptions of uncertainty, while social scientists often employ qualitative assessments. Addressing this divide reflected the increasing engagement of social scientists in the process and a subsequent broadening of the cross-disciplinary approach. The IPCC also dealt with another divide, that between the producers and the recipients of the uncertainty assessments, clarifying the relationship between the qualitative "confidence" and the quantitative "likelihood" languages, where the former is a judgment about the validity of findings, while the latter refers to quantified uncertainty. An intelligence officer would quickly recognize how similar this process was to the intelligence community's methods for assessing sources and their information.

This quantified uncertainty had to be translated from a numerical to a corresponding linguistic scale (see Table 4.1).[66] This exercise, too, would be easily

---

[65] InterAcademy Council, *Climate Change Assessments: Review of the Processes and Procedures of the IPCC* (Amsterdam: InterAcademy Council, 2010). In the Working Group I contribution to the fifth assessment report (September 2013) there was a special section on the treatment of uncertainties and a new and more precise terminology for describing likelihood was introduced. IPCC (2013), Chapter 1, p. 18.

[66] Intergovernmental Panel on Climate Change, *Climate Change 2013 – the Physical Science Basis: Working Group I Contribution to the Fifth Assessment Report of the IPCC (Preliminary Report)* (New York: Cambridge University Press, 2013), pp. 17–18.

Table 4.1 **Likelihood Terms Associated with Outcomes in the Fifth Assessment Report (AR5)**

Term	Likelihood of Outcome
Virtually certain	99–100 percent probability
Very likely	90–100 percent probability
Likely	66–100 percent probability
About as likely as not	33–66 percent probability
Unlikely	0–33 percent probability
Very unlikely	0–10 percent probability
Exceptionally unlikely	0–1 percent probability

recognizable to an intelligence officer. Indeed, recall from Chapter 3 Sherman Kent's charming essay about his effort to do exactly the same thing for the language of national estimates. And US NIEs now come with a similar chart explaining what particular likelihood terms signify.

The IPCC underlined that this whole enterprise in approaching the issue of uncertainty in a more systematic way should be regarded as an interdisciplinary work in progress; it also raised a warning flag that a possible loss of precision in the scheme above could arise when assessments were translated from English to other languages.[67]

Science in an Intelligence Mode?

The three environmental alarms discussed here have more or less obvious similarities with intelligence problems. Intelligence is, as discussed in depth in subsequent chapters, faced with the sometimes unrewarding task of handling assessments of low probability but high impact developments. There is the obvious risk of crying wolf, with inevitable validation if wolves don't appear as predicted, but also the fallacy of being overcautious, either for not wanting to take that risk or for being deterred by the demand for evidence, often not available until too late if at all (sightings of wolves, counting the teeth). Scientists, while in most instances operating far from this exposed position between unfolding events and policymakers unenthusiastic over the delivery of bad news, do in some cases nevertheless share the predicament of intelligence, having to decide at what level of uncertainty a risk should be

[67] Ibid., p. 19.

regarded as corroborated to a degree that it ought to be communicated. Risks are, as a vast sociological and organizational literature illustrates, socially constructed and reconstructed in systems plagued with chronic information overload.[68] Popper's line of demarcation appears in another shape in this warning context—as the dividing line between McCone's 1962 missile hunch, the needle in the haystack of HUMINT from Cuba pointing to a specific site, and the final U-2 aerial photos of the San Cristobal area with the Soviet missile launch facilities in full display. "Boundary work" in these contexts deals with the fluid dividing line between risks and emerging threats and is in some respect what intelligence is all about.[69]

For scientists, the professional role calls for more data and further analysis. Uncertainty is, as vividly illustrated by the IPCC process, first and foremost an incentive for intensified research, for the formulation and testing of new hypotheses. But in cases where broader social values seem to be at stake, this is not a viable position—or is it, after all? In the forest death controversy, not all scientists jumped on the bandwagon, and among those who did, there was a generally recognized need for more detailed surveys, more accurate and comparable data, and research efforts to test the warning hypothesis over time. In principle, the global warming hypothesis of the IPCC has the same scientific foundation, and refined models will gradually be tested against observations, although in a much longer time perspective.

Science in the intelligence mode is illuminating, both for the similarities and the differences compared to national security intelligence. Science is here faced with problems only too familiar to intelligence. One of them is the impact of a dominating paradigm, a concept of the kind that straitjacketed the research department of the Israeli military intelligence prior to the 1973 Yom Kippur War. Since the concept was inadvertently constructed in a way that made refutation impossible, anomalies could be discarded as below the line of demarcation. This tendency was also visible in the acid rain–forest death issue, with the deterring effect of the dominating interpretation of ambiguous observations. In the global warming issue, the risk of a paradigmatic bubble is noted by several observers over time.[70] Dissenting views disappear or are

[68] For a summary of the sociological discourse on uncertainty and risk, see Albert J. Reiss Jr. "The Institutionalization of Risk," in James F. Short Jr., and Lee Clarke, *Organizations, Uncertainties, and Risk* (Boulder, CO: Westview Press, 1992).

[69] The relationship between risk and threat in an intelligence and warning perspective is excellently discussed in Mark Phythian, "Policing Uncertainty: Intelligence, Security and Risk," *Intelligence and National Security* 27, no. 2 (2012): 187–205.

[70] Simon Shackley, "The Intergovernmental Panel on Climate Change: Consensual Knowledge and Global Politics," *Global Environmental Change* 7 (1997): 77–79. See also Tol, "Regulating Knowledge Monopolies."

not properly considered; they tend to be regarded as unnecessary distortion, potentially weakening inherent activism in the IPCC concept and complicating the achievement of the explicit consensus goal.

While the IPCC achieved considerable innovation of the entire research system and the methods, the modus operandi on an aggregated level in the form of collectively negotiated assessments, hybrid science nevertheless encountered problems similar to the pitfalls of intelligence analysis. These included inadvertent establishment of a dominating paradigm that distorted the "boundary work" between risk and threat, the inability to communicate uncertainty in a visible, comprehensible, and actionable way, and finally the very bad track record of consensus intelligence assessments. Intelligence has, as we discussed in this book, much to learn from science and also from science in the intelligence mode, operating in a fluid network structure, interacting with the public, stakeholders, and the policy domain in a way traditionally unthinkable in intelligence. On the other hand, science in the intelligence mode seems to have just as much or even more to learn from intelligence about the classical structural and conceptual roots of failed assessments.

5

Exploring Other Domains

This chapter scans other domains, mostly but not exclusively in the academy, for suggestive ideas to better understand intelligence. In many respects, physicians resemble intelligence analysts: both have limited information and are in the position of conveying judgments that are ultimately subjective to policy officials (or patients) who find it hard to think of probabilities, especially low probabilities of grave consequences. Intelligence typically looks to the methods of social and hard science for rigor, even if it almost never has the opportunity to conduct experiments. But a wide range of other domains are also suggestive. Archaeology, for instance, faces the challenge of very limited data, and journalism confronts the quandary of deciding when an account is validated enough to publish. Even consumer products are suggestive: should intelligence analyses come with a list of ingredients (in the form of methods) and perhaps even a "use by" date?

When one of us (Treverton) used the case of the swine flu pandemic in the United States that never was, in 1976, as the introduction to a week-long executive program for analysts from the CIA and other US intelligence agencies, those analysts immediately saw the doctors as their counterparts. They recognized that the physicians, like them, often were in the position of turning their expertise into essentially subjective judgments about outcomes, some of them very bad but many of them also extremely improbable. Walter Lacquer described this dimension of judgment in positive terms:

> For the truth ("scientific intelligence") is not in the impressive apparatus, the ingenious photographic equipment and the amazing electronic contraptions, it is certainly not in the pseudo-scientific early warning indicators of some students of international relations, but in the small voice of critical judgment, not easy to describe, more difficult to acquire, yet absolutely essential in reaching a correct decision.[1]

[1] Walter Laqueur, "The Question of Judgment: Intelligence and Medicine," *Journal of Contemporary History* 18, no. 4 (1983): 542, 45.

Doctors probably receive less training than intelligence analysts in dealing with probability. Yet both deal with "policy" people, including patients, who are like most humans in being terrible at interpreting probabilities—in the swine flu case, neither were the experts (doctors) very helpful nor were the policy officials very sophisticated. In that, alas, they were both typical. And the hardest cases are those that involve low probabilities of dire outcomes.

In one respect, virtually all professions are similar to intelligence analysis in that they require synthesizing knowledge and drawing inferences. Yet to open the scan that widely would be to produce propositions too general to be useful. Thus, it is helpful to more tightly characterize what might be considered intelligence analysis, or, for that matter, a profession. One of us (Agrell) said it well when he described a profession as requiring the "systematic employment of knowledge, where methods are visible and verifiable, their employment can be tested, and the results can be predicted."[2] Moreover, Agrell aptly characterized intelligence analysis as grappling with a "specific form of uncertainty," the core intellectual requirement being the nature by which this "uncertainty is handled or exploited."[3]

Without wading into the debate as to whether intelligence analysis meets the definition of a profession—and there seems considerable debate—suffice it to say that "unlike many mature professions, intelligence analysis lacks an agreed, unified methodology, and the experts necessary for regulating one."[4] However, Agrell posits that "intelligence analysis combines the dynamics of journalism with the problem solving of science," the combination of which requires the same specific intellectual and psychological qualifications as the experienced news editor or innovative researcher.[5]

Medicine

In many respects, medicine does seem the closest parallel to intelligence. Not only is the analyst-policy relationship much like the doctor-policy (or patient) one; both doctors and analysts are dealing with matters that can, literally, involve life or death. Neither is often in the position of being able to run controlled experiments or even pilot projects. Both are "unscientific" in that neither is much

---

[2] Wilhelm Agrell, "When Everything Is Intelligence, Nothing Is Intelligence," in *Kent Center Occasional Papers* (Washington, DC: Central Intelligence Agency, 2003), p. 3.

[3] Ibid. p. 5.

[4] Matthew Herbert, "The Intelligence Analyst as Epistemologist," *International Journal of Intelligence and CounterIntelligence* 19, no. 4 (2006): 769.

[5] Agrell, "When Everything Is Intelligence, Nothing Is Intelligence," p. 6.

undergirded by usable theory. That is true for medicine despite long histories of work by researchers in relevant disciplines: it may be that the human body isn't much simpler than the universe. In any case, notice how many important drugs were discovered purely empirically, when a drug prescribed for one purpose turned out to have positive effects for quite another malady. As a result, not surprisingly, while more research has looked at various professions to attempt to compare their characteristics to the field of intelligence analysis, the most careful work has been done on medicine, specifically, and social science more generally. This section looks across that work, seeking to capture and summarize the nature of the similarities among a few of the most commonly cited professions.

Not surprisingly given the overlaps, perhaps the most extensive comparative research is that examining intelligence analysis in light of the field of medicine. In 1983, Walter Laqueur observed quite simply that "the student of intelligence will profit more from contemplating the principles of medical diagnosis than immersing himself in any other field."[6] Stephen Marrin, who has dedicated considerable effort to exploring other disciplines for insight into intelligence analysis, perhaps more than any other single researcher, notes that "intelligence analysis is similar to the medical profession in that it requires a combination of skills acquired through practical experiences, and specialized knowledge acquired through academic training."[7]

Moreover, Marrin notes, both fields "use approximations of the scientific method—observation, hypothesis, experimentation, and conclusion— ... to organize and interpret information, and benefit from a variety of technological tools to aid in discernment."[8] Each, nevertheless, "requires critical thinking and judgment to interpret the evidence that goes beyond what can be quantified or automated."[9] His *Improving Intelligence Analysis* synthesizes the research on the similarities of medicine and intelligence; in a 2005 paper on the same subject area, he and Jonathon D. Clemente propose, and justify, that "processes used by the medical profession to ensure diagnostic accuracy may provide specific models for the Intelligence Community to improve accuracy of analytic procedures."[10]

---

[6] Laqueur, "The Question of Judgment." p. 543.

[7] Stephen Marrin, "Intelligence Analysis: Turning a Craft into a Profession," *International Conference on Intelligence Analysis* (2005), p. 1. In Proceedings of the First International Conference on Intelligence Analysis, MITRE, McLean, VA, May, 2005. https://analysis.mitre.org/proceedings/index.html; https://www.e-education.psu.edu/drupal6/files/sgam/IA_Turning_Craft_into_Profession_Marrin.pdf

[8] Ibid., p. 1.

[9] Ibid., p. 2.

[10] Stephen Marrin and Jonathan D. Clemente, "Improving Intelligence Analysis by Looking to the Medical Profession," *International Journal of Intelligence and CounterIntelligence* 18, no. 4 (2005): 1.

They achieve this by working initially from a premise laid down by Laqueur in "The Question of Judgment: Intelligence and Medicine": "The doctor and the analyst have to collect and evaluate the evidence about phenomena frequently not amenable to direct observation. This is done on the basis of indications, signs, symptoms . . . The same approach applies to intelligence."[11] Marrin and Clemente identify an array of similarities with respect to the collection of information, analysis and diagnosis, and the causes of inaccuracy, while also acknowledging the differences and limits of the comparison. In so doing, they demonstrate how analogous subfields of medicine possess characteristics that closely correspond to various intelligence disciplines. For instance, the patient's health history might be thought of an analogous to open source intelligence (OSINT) and the patient interview to human intelligence (HUMINT), then X-rays are the medical equivalent of imagery intelligence (IMINT), an EKG or heart monitor constitutes signals intelligence (SIGINT), and blood tests and urinalysis might be thought of as somewhat like what intelligence calls measurement and signature intelligence (MASINT).[12] In the same way that the patient's health history contributes three-quarters of the information needed for medical diagnosis, OSINT can often contribute at least as large a proportion of the information used to produce an intelligence assessment.[13]

Comparing medical and intelligence "INTs" is more amusing than useful, but comparing functions, say, between general practitioners and intelligence analysts may be more thought provoking. For Marrin, analysts and general practitioners share many roles and functions. The general practitioner has become the all-purpose general diagnostician and subsequent gatekeeper for access to specialists. For intelligence, Matthew Mihelic suggests that the general intelligence analysts can best identify risks and stratify them as to likelihood and potential impact, as does the primary care physician when confronted with a patient who presents few and seemingly disparate symptoms.[14] Moreover, the attributes of evaluation and analysis also correspond closely, namely, forming a hypothesis, performing differential diagnosis, considering symptoms, and finding possible explanations.[15] As Marrin points out,

[11] Laqueur, "The Question of Judgment," p. 535.

[12] Stephen Marrin, *Improving Intelligence Analysis: Bridging the Gap between Scholarship and Practice* (London: Routledge, 2011), p. 108.

[13] Ibid., p. 108.

[14] Ibid., p. 113, and Matthew Mihelic, "Generalist Function in Intelligence Analysis". Proceeding from the 2005 International Conference on Intelligence Analysis, available at https://www.e-education.psu.edu/.../Generalist%20...

[15] Marrin, *Improving Intelligence Analysis*. Also, Stephen Marrin, "Best Analytic Practices from Non-Intelligence Sectors," ed. Analytics Institute (2011).

For the most part, physicians must fit the signs and symptoms together into a hypothesis informed by theory . . . but in both cases ambiguous information and circumstances require critical thinking and judgment in order to come to conclusions regarding the accuracy of the hypothesis and its implication for either national security interests or the patient's well-being.[16]

Yet Laqueur cautions: "No two individuals react alike and behave alike under the abnormal conditions which are known as disease . . . this is the fundamental difficulty in the education of a physician."[17] Likewise, the intelligence analyst faces many, if not more, of the same uncertainties. For this reason, as Marrin suggests, "sometimes intuition is used rather than transparent structured methods visible to outsiders. Assessment in both fields involves the application of cognitive heuristics as convenient shortcuts, which helps achieve accuracy in some cases but can hurt in others."[18]

Deception would seem one area in which medicine would differ sharply from intelligence. Lacquer does note that "the patient usually cooperates with the medical expert." Marrin, however, focuses on similarities behind the apparent differences:

The physician's environment is infused with denial and deception originating from the patients. Sometimes the denial and deception is relatively innocent, as a result of patient lack of knowledge or self-deception. Much of this can result from the patient's decision that the physician does not need to know about particular behaviors related to lifestyle which might be embarrassing to acknowledge. . . . Patients may also hide information related to failure to exercise, take medication, engage in risky behaviors, or consume alcohol, tobacco, or illegal drugs.[19]

Moreover, when a medical patient seeks a particular diagnosis for the purpose of receiving a disability benefit or a prescription narcotic to feed an addiction, medical professionals in those circumstances face challenges that may look more like those faced by their intelligence counterparts when they feel under pressure, directly or not, from policy officials to come up with the

---

[16] Marrin, *Improving Intelligence Analysis*, p. 109.
[17] Laqueur, "The Question of Judgment," p. 536.
[18] Marrin, *Improving Intelligence Analysis*, p. 111.
[19] Ibid., p. 120.

"right" answer.[20] (Chapter 8 further explores this issue, labeled "politicization" in intelligence.)

What is called "evidence-based medicine" has roots going back a hundred years but is newly in vogue. It is suggestive for intelligence, as much in its contrasts as in its similarities to the medical field. It is "the conscientious, explicit and judicious use of current best evidence in making decisions about the care of individual patients."[21] It perhaps proceeds from the old adage that when medical students hear hoofbeats, they diagnose zebras, when virtually all the time those beats will be horses; high temperatures and body aches almost always signify colds or flu. In that sense, evidence-based medicine, which explicitly accepts the value of a doctor's "individual clinical expertise," seeks to ground that expertise in evidence. In that it is unlike intelligence, which virtually always begins with evidence. It may be unlike intelligence in that most of the time new cases mirror existing evidence: those hoofbeats are usually horses. Evidence-based medicine is also most powerful in prescribing treatments, where the evidence of what works is likely to be richer. Thus, "protocols" specify which treatment the evidence indicates is best for a particular ailment, putting the onus on any doctor who recommends something else to specify why.

Social Science

Intelligence analysts bear strong similarity to social scientists when they create, evaluate, and test hypotheses as part of a rigorous, structured approach to analysis.[22] Indeed, comparison between intelligence analysis and social science has a long legacy. It was perhaps Sherman Kent who first noted the similarity, at least in the American context of a formalized intelligence structure, in 1949: "most of the subject matter of intelligence falls in the field of the social sciences."[23] Conceptual models used in the deductive process of intelligence analysis derive from what Roger Hilsman described as "the way that the social sciences adapt the scientific method to create and test hypotheses . . . to derive meaning from the accumulated data."[24] Washington Platt echoed these sentiments, noting the extent to which intelligence analysis could be improved by relying on the social

---

[20] Marrin, "Best Analytic Practices from Non-Intelligence Sectors."

[21] See David L. Sackett, "Evidence-Based Medicine," *Seminars in Perinatology* 21, no. 1 (1997): 3–5.

[22] Marrin, "Best Analytic Practices from Non-Intelligence Sectors."

[23] Kent, *Strategic Intelligence for American World Policy* (Princeton, NJ: Princeton University Press, 1949), p. 175.

[24] Marrin, *Improving Intelligence Analysis*, p. 24, originally from Roger Hilsman, "Intelligence and Policy-Making in Foreign Affairs," *World Politics* 5 (1952): 1–45.

sciences; he suggested that "In nearly all problems confronting the intelligence officer, some help, even if not necessarily a complete answer, is available from those who have already wrestled with similar questions."[25]

In *Improving Intelligence Analysis,* Marrin too addresses the trajectory of social science and its applicability to intelligence analysis and argues that "intelligence analysis is rooted in the methodologies and epistemologies of the social sciences."[26] For Marrin, it was Klaus Knorr who best evaluated how analysts use social science. Knorr's view held that "the central task of the intelligence officer, historian, and social scientist is to fit facts into meaningful patterns, thus establishing their relevance and bearing to the problem at hand."[27] Moreover, Knorr's account of the suitability of "social science methods of gathering data, of deducing data from other data, and of establishing the validity of data that are of particular value . . . in producing appropriate kinds of information for intelligence" is apt.[28]

Through the years, a series of initiatives have sought to impel intelligence analysis to rely more, and more systematically, on the best of social science methods. For instance, in 1999, the special assistant for intelligence programs at the US National Security Council, Mary McCarthy, suggested that academia develop an arsenal of social scientific methodologies—later termed "structural analytic techniques"—so that analysts could "understand how thinking can be done and how methodologies work."[29] Using formal methods had the potential to enable an analytic audit trail, through which "analysts and their colleagues can discover the sources of analytic mistakes when they occur and evaluate new methods or new applications of old methods."[30]

As both the study and practice of intelligence analysis incorporates more explicit reference to social science methodology, greater emphasis is being placed on structured analytical techniques as a mechanism for embedding social science methodologies within analytic practices. Optimally, structured analytic techniques provide analysts with simplified frameworks for doing analysis, consisting of a checklist derived from best practices in application of the scientific method to social issues.[31] More recently, a major National

[25] W. Platt, *Strategic Intelligence Production: Basic Principles* (New York: F. A. Praeger, 1957), p. 133.

[26] Marrin, *Improving Intelligence Analysis,* p. 28.

[27] K. E. Knorr, *Foreign Intelligence and the Social Sciences* (Princeton, NJ: Center of International Studies, Woodrow Wilson School of Public and International Affairs, Princeton University, 1964), p. 23.

[28] Ibid., p. 11.

[29] Marrin, *Improving Intelligence Analysis,* p. 31.

[30] Ibid.

[31] Ibid., p. 32.

Academy of Sciences study in the United States concluded that the best way to improve intelligence analysis would be to make more, and more systematic and self-conscious, use of the methods of social science.[32]

To be sure, social scientists differ, and so do their methods. When the intelligence task is piecing together fragmentary information to construct possible explanations under given circumstances, intelligence analysis appears very similar to the work of historians.[33] Historians describe and explain while crafting a portrayal of the past. Yet they work from an archive as their source material, which is synthesized to support a perspective. Historian John Lewis Gaddis compares history to geology, paleontology, and astronomy because the scientists are unable to directly evaluate their subject matter, just as is the case in intelligence analysis. Moreover, in recognizing this, some observers have asked, "What can we learn from how academic history is done, with respect to the field of intelligence?" Can evaluating the quality of history (as a social science discipline) suggest lessons for evaluating the quality of intelligence?[34]

What most obviously distinguishes intelligence from social science is, as discussed in Chapter 2, the institutional context. Intelligence analysis as produced in governments has an active audience of policy officials looking for help in making decisions. It also has access to information that was collected secretly, including ways that are illegal in the jurisdictions in which it occurred. And because intelligence is seeking to give one nation a leg up in international competition, it is subject to deception from the other side in a way social science is not. Those from whom social scientists seek information—in a survey instrument, for example—may be reluctant or too easily swayed by fashion in what they "ought to" think. They are probably less likely, however, to tell outright lies, at least beyond those questions that might reveal behavior that was embarrassing or illegal—and for those questions the social scientist is forewarned of unreliability.

Yet purpose is perhaps a greater distinguisher of intelligence from social science. Not only does intelligence have an interested audience, in principle at least, but it also usually is required to be *predictive*. Explanation may be enough for social science, and it is fundamental to intelligence analysis as well. But it is usually not enough for the policy officials who need to act. Stéphane Lefebvre offers a description of intelligence analysis that could, institutional context aside, characterize social science as well:

---

[32] Committee on Behavioral and Social Science Research to Improve Intelligence Analysis for National Security, National Research Council, *Intelligence Analysis for Tomorrow: Advances from the Behavioral and Social Sciences* (National Academies Press, 2011).

[33] Marrin, "Best Analytic Practices from Non-Intelligence Sectors."

[34] Ibid.

Broadly understood, intelligence analysis is the process of evaluating and transforming raw data acquired covertly into descriptions, explanations, and judgments for policy consumers. It involves assessing the reliability and credibility of the data, and comparing it with the knowledge base available to the analyst, to separate fact from error and uncover deception.[35]

Yet Lefebvre continues to highlight prediction as a fundamental difference: "Most intelligence analysis is predictive in nature and follows a simple pattern: it describes what is known, highlights the interrelationships that form the basis for the judgments; and offers a forecast."[36] The predictive nature of intelligence is a fundamental differentiator and a practical requirement, not needed or necessarily practiced in the social sciences. Intelligence has no luxury of just explaining; the possibility of a terror plot at hand casts the judgment used to evaluate intelligence information in a fundamentally different light from that used in the social sciences. As with medical doctors, the "intimate connection with life and death" both makes it imperative to make a call and lends caution to the analysis of intelligence in a national security context.[37] For this reason, the fear of failing to predict an attack might lead to an analytic "threat bias, somewhat similar to the "wet bias" of weather forecasters. Nate Silver points out:

Meteorologists at the Weather Channel will fudge a little bit under certain conditions. Historically, for instance, when they say there is a 20 percent chance of rain, it has actually only rained about 5 percent of the time. . . . People notice one type of mistake—the failure to predict rain—more than another kind, false alarms. If it rains when it isn't supposed to, they curse the weatherman for ruining their picnic, whereas an unexpectedly sunny day is taken as a serendipitous bonus. . . . Their wet bias is limited to slightly exaggerating the probability of rain when it is unlikely to occur—saying there is a 20 percent chance when they know it is really a 5 or 10 percent chance—covering their butts in the case of an unexpected sprinkle.[38]

[35] Stéphane Lefebvre, "A Look at Intelligence Analysis," *International Journal of Intelligence and CounterIntelligence* 17, no. 2 (2004): 236.

[36] Ibid.

[37] N. Silver, *The Signal and the Noise: Why So Many Predictions Fail–but Some Don't* (New York: Penguin Group, 2012), Location 3920, Kindle Edition.

[38] Ibid., Location 2355–63, Kindle Edition. This approach is somewhat similar to the "minimum expected chagrin" explained by Alvan Feinstein, "The 'Chagrin Factor' and Qualitative Decision Analysis," *Archives of Internal Medicine* 145, no. 7 (1985).

By contrast to intelligence analysts and weather forecasters, social scientists have the luxury of waiting. It is in the nature of the scientific enterprise that claims are not made about the existence of phenomena unless a high level of confidence can be asserted for such claims. A 95 percent confidence level is a typical cut-off point: Scientists defer making claims or predictions until they can be highly confident in them. In short, science requires a high burden of proof for concluding that a climate phenomenon exists or that a projection of future climate conditions—especially extreme conditions—is solidly supported.

In security policy the practice for deciding whether to take a hazard seriously is much different from the practice in making scientific claims. Security analysts are focused on risk, which is usually understood to be the likelihood of an event multiplied by the seriousness of the consequences if it should occur. Thus security analysts become seriously concerned about very high-consequence negative events even if scientists cannot estimate the probability of their occurrence with confidence and, indeed, sometimes even if they are fairly confident that the probability is quite low.[39]

To be sure, the differences between intelligence analysis and social science are considerable. Academic social scientists usually have considerable discretion in the problems they address. Not so, intelligence analysts who, ideally, are meant to be in the position of responding to requests from policy officials, though more often in fact they are addressing issues their office *thinks* are on the agenda. In either case, they lack the discretion of academics. So, too, do they usually lack the leisure of time; often, the answer is desired yesterday. The last difference is probably also the biggest, and it is one that runs through this book. Academic social scientists have robust data sets much more often than intelligence analysts; indeed, given discretion, the social scientists can choose topics on which they know there are data. By contrast, intelligence analysts live in a world of spotty data, usually collected opportunistically; it is likely to amount to a biased sample, though biased in ways that analysts may not know, making it impossible to judge how much the sample can be applied more generally to a broader population. As a result, intelligence analysts generally cannot employ the methods of social science in a robust way.[40]

In "The Intelligence Analyst as Social Scientist: A Comparison of Research Methods," Henry W. Prunckun Jr. offers an intriguing account of the differences

[39] John D. Steinbruner, Paul C. Stern, and Jo L. Husbands, *Climate and Social Stress: Implications for Security Analysis* (National Academies Press, 2013).

[40] Marrin, "Best Analytic Practices from Non-Intelligence Sectors."

and similarities from an unusual perspective, that of law enforcement intelligence analysis. The study provides a short comparison of qualitative collection and practical analytic methods used by both intelligence analysts and social scientists.[41] In the way it is organized, law enforcement resembles social science, as well as intelligence beyond law enforcement. The law enforcement function typically comprises two groups—one for support, the other for intelligence. The latter typically is divided in two, between tactical and strategic. The strategic analysis is looking at longer-term trends in criminality, trying to identify how law enforcement resources should be deployed while also pointing toward emerging threat. The former is much more operational day-to-day. That tactical function might be likened to project teams in government, industry, or other private sector organizations, conducting research looking more specifically at how particular programs are performing.

On the collection side, law enforcement is constrained mostly by law, while foreign intelligence services often are less constrained by law. For law enforcement, informants, wiretaps, and other forms of surveillance are only suggestive of the tools, while foreign intelligence may be even less constrained in employing kin of those tools, at least outside the agency's home country. However, social science research is, in another sense, less constrained: it can choose its hypotheses to test and can run experiments or other surveys to test them. For it, the main constraints are ethical about how it treats those who are the subjects of their research in one way or another. In some respects, though, the differences may be less than meets the eye: observation by field researchers in social science may look a lot like informants and observation in criminal intelligence, minus the gadgetry.

If social science research typically begins with a hypothesis, law enforcement intelligence usually starts with a target—individuals or groups of concern because they are engaged in criminality or thought to be likely to do so. Its process is not very different from the canonical intelligence cycle, from devising a plan to collect relevant data through collecting it, to analysis and dissemination. Especially on the tactical side, the process is dynamic: policing resources may shift, requiring an examination of whether the target in question is still a threat. In actually doing the analysis, law enforcement employs, if sometimes fairly crudely, a range of techniques from social science. The vast majority of its data is qualitative—that is, unstructured—and so its "association charts" of would-be criminals and their affiliates is easily recognizable to social scientists as "network analysis."[42] So, too, both social science and police

---

[41] Henry Prunckun, "The Intelligence Analyst as Social Scientist: A Comparison of Research Methods," *Police Studies* 19, no. 3 (1996).

[42] Ibid., p. 73.

intelligence make use of content analysis—for instance, published sources, interview transcripts or focus group session for the former, with the latter adding conversations gathered by electronic surveillance. Biographic analysis is common to both enterprises, while financial analysis is more common in police intelligence. In both cases, advances in dealing with large data sets make it more possible to search for and detect patterns in unstructured data.

Yet part of the culture of analysis and its separation from policy, again especially in the United States, may reflect a kind of envy of both social and natural sciences. Analysts may aspire, perhaps not even entirely consciously, to be ideal-type policy experts as they perceive it from the sciences. In Bruce Bimber's words,

> [An] idealized image of the scientific expert involves not simply knowledge, but also a large element of objectivity, of being above politics and partisanship. The idealized policy expert brings the neutral authority of science to bear on politics. Experts derive legitimacy from their ability to appeal to non-political political standards: the use of dispassionate scientific methods of inquiry, validation through peer review rather than mere assertion, and other classic elements of . . . science.[43]

Anthropology and Sociology

Exhortations to make better use of social science are not likely to be music to the ears of intelligence officers, either because they seem too mechanical[44] or because they seem quaintly unaware of the pressures, including political, under which intelligence works.[45] Social science often seeks regularities in human behavior, while intelligence often is looking for particularities. Intelligence professionals have a tendency to view any particular situation as specific to the precise context in which it occurs, and intricately connected to the personalities involved; thus, for their purpose, it is a sample size of one. As Lacquer put it: "the more distinctly human the phenomenon, the more necessary a human

[43] Bruce Bimber, *The Politics of Expertise in Congress: The Rise and Fall of the Office of Technology Assessment* (Albany: State University of New York Press, 1996), p. 12.

[44] For instance, the recommendations of the US National Academy's Committee on Behavioral and Social Science Research to Improve Intelligence Analysis for National Security do not seem to go much beyond identifying social science methods, particularly more quantitative ones, that might be relevant to intelligence. Nor do the recommendations pay attention to the context in which these methods are to be employed.

[45] Fulton Armstrong, "The CIA and WMDs: The Damning Evidence," *New York Review of Books*, 2013, http://www.nybooks.com/articles/archives/2010/aug/19/cia–and–wmds–damning–evidence/?pagination=false.

observer."[46] History also views circumstances as unique, but since it is about the past, it can be a component of intelligence analysis but not a comparable guide to method.

When the intelligence task is attempting to see the world through the eyes of another culture, using knowledge, insight, or other understanding of this nature, it can appear very similar to the work of an anthropologist.[47] Thus, it is no surprise that intelligence agencies are rediscovering anthropologists once again. In the US case, that has been driven mostly by the challenges of fighting wars, hence the odd label "human terrain awareness."[48] In one 2007 case in southern Afghanistan, coalition forces encountered a barrage of suicide attacks. One anthropologist working with the military identified an unusually high concentration of widows in poverty in the region, creating pressure on their sons to join the well-paid insurgents. Following that advice, US officers developed a job training program for the widows.[49]

As observed by one of the creators of human terrain analysis, Montgomery McFate: "one of the central epistemological tenets of anthropology is cultural relativism—understanding other societies from within their own framework."[50] Similarly, in 1986 Bernard and colleagues noted that "Anthropologists are increasingly aware ... of the fact that every field situation is different and can never be duplicated."[51] Yet much of the description surrounding anthropology, particularly as it has been applied in the context of national security, describes its value in terms of understanding the cultural landscape; moreover, anthropology views the human terrain, so to speak, "not as something to be controlled, but as part of the environment that must be understood ... to succeed."[52] The emergence of terrorism as a primary focus of intelligence in the early 2000s has given it a tactical orientation, one different from the orientation during the Cold War, and that has colored the use of anthropology.

[46] Laqueur, "The Question of Judgment: Intelligence and Medicine," pp. 542, 45.

[47] Marrin, "Best Analytic Practices from Non-Intelligence Sectors."

[48] For an assessment of the method in Iraq and Afghanistan, see Montgomery McFate and Steve Fondacaro, "Reflections on the Human Terrain System During the First 4 Years," *Prisms* 2, no. 4 (2011).

[49] As reported in David Rohde, "Army Enlists Anthropology in War Zones," *New York Times* (2007), http://www.nytimes.com/2007/10/05/world/asia/05afghan.html?incamp=article_popular_4&pagewanted=all&_r=0.

[50] Montgomery McFate, "Anthropology and Counterinsurgency: The Strange Story of Their Curious Relationship," *Military Review* 85, no. 2 (2005): (pages not marked).

[51] H. Russell Bernard et al., "The Construction of Primary Data in Cultural Anthropology," *Current Anthropology* 27, no. 4 (1986): 383.

[52] Montgomery McFate and Steve Fondacaro, "Cultural Knowledge and Common Sense," *Anthropology Today* 24, no. 1 (2008): 27.

There is nothing new about intelligence turning to anthropology. For instance, in the 19th century, anthropologists, along with archaeologists and geographers, were used by—and the profession tainted through association with—colonial offices in Britain and France.[53]

Anthropologists played an influential role in the world wars of the last century, and McFate suggests that anthropologist Gregory Bateson conceived the idea to "establish a third agency to employ clandestine operations, economic controls, and psychological pressures in the new warfare"—an idea that later became the Central Intelligence Agency.[54] The cloak of anthropology long has provided a useful cover story for espionage.[55] However, as British and French experiences suggested, from the beginning the intelligence-anthropology interface has been far from uncomplicated, highlighting the underlying ethical and professional problems of the intelligence-academy divide. For instance, Project Camelot, was a US Army research project begun in 1964 but canceled after congressional hearings in 1965. Led by a Chilean-born anthropologist, the goal of the project was to assess the causes of conflict between national groups in order to anticipate social breakdown and eventually suggest solutions. The project was very controversial among social scientists, many of whom voiced concerns that such a study was in conflict with their professional ethics.[56]

At the level of what intelligence would call collection and anthropology "fieldwork," the two share a focus on individuals or small groups and the purpose of enhancing understanding of thoughts, actions, and relationships to groups, subgroups, and other social structures.[57] As H. Russell Bernard says with respect to anthropology, "all data gathering in fieldwork boils down to two broad kinds of activities: watching and listening. . . . Most data collection

[53] For a case study of pre–World War I employment of a Western explorer, see David W. J. Gill, "Harry Pirie-Gordon: Historical Research, Journalism and Intelligence Gathering in Eastern Mediterranean (1908–18)," *Intelligence and National Security* 21, no. 6 (December 2006): 1045–1059.

[54] McFate, "Anthropology and Counterinsurgency," page not marked. This claim is originally sourced to Arthur Darling, "The Birth of Central Intelligence," Sherman Kent Center for the Study of Intelligence, online at www.cia.gov/csi/kent_csi/docs/v10i2a01p_0001.htm.

[55] Ibid., pages not marked. Note: McFate's telling of this narrative and the relationship between anthropologists and the national security establishment received criticism: Jeffrey A. Sluka, "Curiouser and Curiouser: Montgomery Mcfate's Strange Interpretation of the Relationship between Anthropology and Counterinsurgency," *PoLar: Political and Legal Anthropology Review* 33, no. s1 (2010): 99–115.

[56] For a once famous example of this controversy, see *The Rise and Fall of Project Camelot. Studies in the Relationship between Social Science and Pratical Politics*, ed. Irving Louis Horowitz (Cambridge, MA: MIT Press, 1967).

[57] H. Russell Bernard, *Handbook of Methods in Cultural Anthropology* (Walnut Creek, CA: AltaMira Press 1998), p. 313.

in anthropology is done by just talking to people."[58] Human intelligence could be likewise described. One difference, though, is that while the business of intelligence traditionally separated collection from analysis, an anthropologist is both field collector *and* analyst. That said, anthropology's four main types of primary data construction are very much akin to intelligence:[59]

1. Unstructured interviewing with key informants
2. Structured interviews of respondents
3. Direct observation
4. Archival data

Unstructured interviewing in the social sciences is a rough parallel to the operational meetings of the human intelligence case officer.[60] Intelligence officers have specific questions or themes to address, and clear objectives for each meeting, though the source is not prevented from drifting . . . so long as the information maintains relevance. Unstructured interviewing also helps in the early stages of building rapport and can ultimately move into sensitive topics, for both disciplines.[61] Rapport building (sometimes also considered "impression management") is a cornerstone skill of human intelligence and a critical element of anthropological research. H. Russell Bernard admits, however, that the definition of rapport is elusive, but this state is achieved when informant and researcher come to share the same goals.[62] Physical presence ("hanging out") builds trust, or rapport, in the anthropological sense.[63] Anthropology and intelligence both target activities or settings to acquire an enhanced understanding of the environment.[64] Acquisition of information for both HUMINT and anthropology is dependent on the access proximity of "sources" having the desired information, and their disposition and perspective on the subject matter.

The study and evaluation of human behavior and interaction as well as social groupings, through elicitation, observation, and the scientific method, are the work of sociologists and anthropologists, but they are also the work

---

[58] H. Russell Bernard, *Research Methods in Anthropology: Qualitative and Quantitative Approaches* (Walnut Creek, CA: AltaMira Press, 2006), p. ix.

[59] Bernard et al., "The Construction of Primary Data in Cultural Anthropology," *Current Anthropology* 27, no. 4 (1986): 382.

[60] Bernard, *Research Methods in Anthropology*, p. 211.

[61] Ibid., p. 213.

[62] Bernard, *Handbook of Methods in Cultural Anthropology*, p. 268.

[63] Bernard, *Research Methods in Anthropology*, pp. 354, 68, 69.

[64] Bernard, *Handbook of Methods in Cultural Anthropology*, p. 314. In the field of intelligence, though, the collection method chosen is that which is most likely to reveal/acquire the desired information.

of the intelligence analyst and case officer. While these similarities exist, intelligence and anthropology diverge in important ways; the two face different constraints. For instance, anthropologists are constrained by the ethical guidelines of the profession. These scientists are meant to observe and understand with as little disruption or influence on their subjects as possible. By contrast, HUMINT case officers, in particular, are both taught and encouraged to manipulate situations, people, and their environment to acquire information; moreover, they often provide monetary compensation to sources in exchange for cooperation. While aspects of intelligence might be considered a subset of anthropology, intelligence employs the practice of "tasking" a source to acquire specific information, which is not a characteristic of anthropology. A similarity is that anthropological informants might desire anonymity and confidentiality in the interaction with anthropologists, just as human sources usually want to hide their association with intelligence from public knowledge.[65]

Moreover, traditional anthropological research is often guided by the methodological stricture to acquire a representative sample of the target population and may seek to generalize findings across the population/subgroup. Anthropologists thus often choose informants who seem representative, or can provide a "grand tour" of the cultural landscape but can and will incorporate the perspective of an isolated group member who maintains the perch of an outsider.[66] By contrast, intelligence analysts typically work with information that is from potentially a biased sample, frequently acquired opportunistically, with no way to determine the generalizability to the broader population, and often with no particular interest in doing so.[67] The collection of intelligence is principally driven by targeting people or groups who can provide privy information on the actions of specific individuals, events, or other information, and the likelihood of specific decisions or actions among these individuals. Targeted intelligence sources might be considered key informants in the anthropological sense, though the primary selection criteria are access to information within a particular context.[68] The degree of cooperation and the information to which the source has access will determine what can actually be collected.

In sum, the defining criterion for a target for intelligence is that target individual's access to specific information needed to fill a gap in knowledge. With the increase in "privy-ness" of information comes less

[65] Bernard, *Research Methods in Anthropology*, p. 315.
[66] Bernard, *Handbook of Methods in Cultural Anthropology*, p. 313.
[67] Marrin, "Best Analytic Practices from Non-Intelligence Sectors."
[68] Bernard, *Research Methods in Anthropology*, p. 196.

certainty in evaluating its credibility, and thus greater risk when making judgments based upon that information.[69] As with other social sciences, anthropological research often seeks representativeness (generalizability), and might be doctrinally inclined to defer to community consensus or formal tests of validity, whereas the world of intelligence is willing to entertain the notion that there is such a thing as information so good it can't be verified. Because the nature of intelligence information desired in many instances is specific with respect to terror organizations or privy information, it is available from only a small population and an even smaller sample who might willingly reveal information should they have access to it. Sources are targeted specifically because the intelligence organization believes the source has access to the desired information. Anthropology can face the same dilemma; H. Russell Bernard indicates that "In depth research on sensitive topics requires nonprobability sampling."[70] However, non-probability sampling leads to potential bias and collects information from a sample that provides only a limited perspective.

In this respect, both intelligence and anthropology struggle with the absence of a worthy mechanism for verifying credibility, informant accuracy, and informant reliability.[71] The nature of this problem is confounded for intelligence, which sees a wide variety of motivations for the cooperation it receives from sources. Anthropology faces is own variety of other problems.[72] Knowledge tests or falsification tests can help to assess the state of knowledge of the source or respondent.[73] For intelligence, the motivations of informants range from patriotism and ethical views, to money, revenge, intent to deceive, or any number of other reasons. The accuracy of information provided by different informants will differ widely. Access is a critical difference for intelligence. So is deception,

[69] An article in *The New Yorker* magazine regarding the prosecution of war crimes in Kosovo provides an excellent example of the type of investigation and analysis that an intelligence operator or analyst might confront; moreover, it also provides an excellent example of why quantitative techniques are often not useful in this context. Nicholas Schmidle, "Are Kosovo's Leaders Guilty of War Crimes?" *The New Yorker,* May 6, 2013.

[70] Bernard, *Research Methods in Anthropology*, pp. 186, 217–222. Also, a note on elicitation: Depending on the nature of the relationship (in an intelligence and anthropological context), and how well developed it might be, intelligence personnel and anthropological people might both use a variety of probes to get the source to provide more information or context about a particular matter of interest. Additionally, both will need to be capable of directing a conversation gracefully toward the issues of interest.

[71] An element of this mindset seems to be present in the Bob Drogin's telling of the story with respect to the intelligence reporting regarding the presence of WMD in Iraq. Bob Drogin, *Curveball: Spies, Lies, and the Con Man Who Caused a War* (New York: Random House, 2007).

[72] Bernard, *Research Methods in Anthropology*, p. 245.

[73] Bernard, *Handbook of Methods in Cultural Anthropology*, p. 379.

though informants in anthropology sometimes lie, too.[74] Similarly, anthropologists, like other social scientists using survey data, must also account for the distinction between what "informants may say and what may be the actual custom of the primitive society"—something like the difference between how survey respondents think they should respond and what they really think.[75]

This distinction strikes at another fundamental differentiator between intelligence and anthropology. Conclusions drawn by social scientists, anthropologists included, often only capture attitude, opinions, norms, values, anticipation, or recall.[76] Policymakers, and thus intelligence officers, are interested in none of these things: they are interested in overt behavior.[77] Sociology seems to have struggled, or is perhaps still struggling, with these inherent problems; Deustcher points out that despite little contention that "what men say and what they do are not always in concordance," it has "no effect at all on the sociological faith in the value of data gathered via opinion, attitude, and other kinds of questionnaires."[78] However, activity-based intelligence, or ABI, as discussed in Chapter 9, could narrow the difference by making intelligence more dependent on understanding attitudes, opinions, and norms in order to comprehend behavior.

Indeed, anthropologists' bread and butter is studying and evaluating human behavior and unpredictability, social groupings, and interaction, using elicitation, observation, and the scientific method in an attempt to explain. This is also the work of sociologists and psychologists, but anthropologists bring a strong sense of the "otherness" of those they study, and so in making their observations they may be more aware of the subtle ways in which the assumptions, even the norms, of their own culture may introduce a bias. What they bring to the task is similar to what is required of both spy-masters and analysts in intelligence—the ability to see through the eyes of those they study, and imagine what it would be like "to walk in their shoes," a critical dimension in forming judgments about foreign events. In process, if not always in purpose or targets

---

[74] Bernard, *Research Methods in Anthropology*, p. 200. In a separate article, Bernard et al. concluded from a survey of research regarding informant accuracy that "on average, about half of what informants report is probably incorrect in some way." Also, H. Russell Bernard et al., "The Problem of Informant Accuracy: The Validity of Retrospective Data," *Annual Review of Anthropology* 13 (1984): 495–517.

[75] Irwin Deutscher, "Words and Deeds: Social Science and Social Policy," *Social Problems* 13(1965): 236.

[76] Ibid., p. 235.

[77] Ibid.

[78] Ibid., p. 237.

Table 5.1 **Comparing Intelligence and Anthropology**

Intelligence Cycle	Human Intelligence	Anthropology
Planning	• Identification of information requirement • Background Research • Targeting	• Identify Problem & Sample Population • Literature Review • Research Proposal
Collection	• Human Source Operations • Rapport Building	• Unstructured Interviewing • Rapport Building
Processing	• Meeting Reports • Grading & Evaluation of Information • Assessment of Credibility	• Field Notes • Capture the nature of the information, and factors that might discredit or support the information provided.
Analysis	• Synthesis • Interpretation	• Synthesis • Interpretation
Dissemination	• Release information to wider community for "consumption"	• Publication: release information to wider community

or constraints, intelligence and anthropology look quite similar, as Table 5.1 indicates.

Journalism

When intelligence analysts explain the facts, driving factors, and potential outcomes of a fluid situation, a current intelligence briefing can appear similar to what journalists do.[79] When one of us (Treverton) was running a workshop for US intelligence analysts, he invited a distinguished correspondent from the Cable News Network (CNN) to speak. The correspondent began by reversing the usual caveats: please protect me, he said, because you and I are in the same business, and we run the same dangers. Indeed, in many parts of the world, they think I *am* you, an intelligence operative. The production of current intelligence and the corresponding skills necessary to skillfully evaluate the significance of an event, and communicate it with appropriate accuracy, is in many respects

---

[79] Marrin, "Best Analytic Practices from Non-Intelligence Sectors."

a journalistic endeavor.[80] The similarity between journalism and intelligence is readily apparent to anyone who visits most intelligence agencies: they have the feel of a newsroom much more than an academic department. The similarities between intelligence and journalism were recognized in the earliest days of intelligence postwar analysis. In 1949, Sherman Kent stated a belief that intelligence organizations needed many of the same qualities as the greatest metropolitan newspapers, and drew direct comparison for intelligence to their products, the fundamental skills required, means of operation, and standards.[81]

When one of us (Treverton) managed the US national intelligence officers in the 1990s, he tried to hire a practicing journalist as an National Intelligence Officer (NIO). He didn't succeed, but the idea was to bring into intelligence the example of someone who didn't turn immediately to the classified computer with all of intelligence's secret sources when something happened, but rather began to work the telephone and not to confine his calls within the United States. More recently, the head of intelligence for the NATO coalition in Afghanistan, ISAF (International Security Assistance Force), General Michael Flynn, proposed that intelligence analysts assigned to the coalition should behave more like journalists, traveling frequently from headquarters to lower-level units—in effect becoming their own collectors.[82]

While medicine on an analytic level resembles much of the assessment problem encountered in intelligence—including that of conveying sometimes unwanted and frightening probabilities—journalism is thus in a broader sense the profession closest to intelligence, sometimes too close in the opinion of either of the two—or both. As the CNN-comment reflects, intelligence operatives and journalists not only cover overlapping events but also operate in the same physical and human terrain. A number of professionals are crossing each other's paths more or less literally: soldiers, diplomats, NGO employees, journalists, and intelligence officers—the last, however, not always with their affiliation on their business card but rather posing as one of the former, most often a journalist.

In their book, *Spinning Intelligence: Why Intelligence Needs the Media, Why the Media Needs Intelligence,* Robert Dover and Michael S. Goodman describes the two professions as "blood brothers" though separated at birth.[83] Intelligence has, since

---

[80] Loch Johnson's article nearly three decades ago addressed the CIA's use and misuse of journalists. Along the way, though, it drives home how kindred intelligence and journalism are. See his "The CIA and the Media," *Intelligence and National Security* 1 (2) May 1986: 143–169.

[81] Kent, *Strategic Intelligence for American World Policy.*

[82] Michael Flynn, *Fixing Intel: A Blueprint for Making Intelligence Relevant in Afghanistan* (Washington: Center for a New American Century 2010).

[83] Robert Dover and Michael S. Goodman, eds., *Spinning Intelligence. Why Intelligence Needs the Media, Why the Media Needs Intelligence* (New York: Columbia University Press, 2009), p. 7.

the advent of modern media, used journalists as sources, informers, and covers but also as a professional pool for recruitment, based on the fact that journalistic skills are so similar to those intelligence usually seeks. Just how close the skills are is perhaps best illustrated by one of the 20th century's best-known intelligence officers and spies, Kim Philby, recruited by the British intelligence after proving his ability as a journalist covering the Spanish civil war on the Franco side (incidentally encouraged to do so by his NKVD controllers in order to become a suitable candidate for recruitment on the Secret Intelligence Service, the SIS).[84] As an intelligence officer, and as an agent for the KGB, the Soviet intelligence agency, Philby displayed a remarkable ability to socialize, build networks, and seize opportunities, in much the same way an experienced foreign correspondent would operate. But Philby also commanded the journalistic core competence, that of reporting. Some of his written reports to his Soviet masters have emerged out of the former KGB archive and constitutes if not the best so at least the best-written and most comprehensible description of the British wartime intelligence labyrinth.[85]

The "blood brothers" are however not always on cooperative or even speaking terms. Intelligence tends to regard media output as incomplete, shallow, or simply flawed. Media have, in their view, their own agenda, which is true—but so has intelligence as we discuss in Chapter 8. It is also true that media usually gives pride of place to rapid dissemination rather than in-depth analysis and long-term assessments. It is very much a real-time activity. Yet so, often, is current intelligence, whether in an operational context in a conflict zone or in various forms of fusion center. The main difference is of course the purpose, but core competences also differ; in the media world, importance is on the ability of journalists to sense and produce a good story without any undue delay, or if necessary in real-time. In some respects, news media are so efficient that intelligence simply use them as an information baseline, especially in rapidly evolving situations in a physical space where media have superior coverage and can rely on crowd-sourcing.[86] The advent of the 24-hour news cycle has, according to Dover and Goodman, moved intelligence agencies on both side of the Atlantic closer to media outlets, not only in the sense of updating, but also in terms of managing their own relations with the public.[87]

[84] On the recruitment of Philby, see Nigel West and Oleg Tsarev, *The Crown Jewels* (London: HarperCollins, 1999).

[85] West and Tsarev, *The Crown Jewels,* Appendix II, "The Philby Reports."

[86] One example of this reliance is the aftermath of the July 22, 2011, terror attacks in Norway, when the staff at the Norwegian Joint Headquarters in Bodø had to rely on live reporting on national television to update on the evolving events. See Wilhelm Agrell, *The Black Swan and Its Opponents. Early Warning Aspects of the Norway Attacks on 22 July 2011* (Stockholm: National Defence College, Centre for Asymmetric Threats, 2013), p. 53.

[87] Dover and Goodman, *Spinning Intelligence,* p. 9.

Three aspects of the journalists' profession stand out as especially relevant for intelligence from a methodological perspective:

- *Integrated information gathering–sensemaking–storytelling loop.* While in many ways similar to an intelligence process, the media process nevertheless features, by virtue of necessity, a number of integrations and shortcuts. While teams out on the spot and the editors at the news desk can function in much the same way as an analytic staff and collection agencies, editors often conduct their own parallel information gathering, while the teams or correspondents can function in their own independent collection/analysis/editing process, providing the news desk at home with the final product ready to send or print. Journalists have here been forced to command a skill certainly not common among, for instance, social scientists (or intelligence analysts), that of being brief, precise, trustworthy—and engaging.
- *Ability to spot news.* The journalistic process is truly Clausewitzian; journalistic work cannot, beyond a certain point, be foreseen, planned, and conducted in a systematic way. This means that journalists on the whole are far better than intelligence analysts at understanding and reacting to new information. Temptations to defend the ramparts of a paradigm, so problematic in science and intelligence analysis, are lower in journalism; journalists are intent on producing stories, unlike scientists who are building theories or intelligence analysts defending assessments. The price of being wrong is also lower for journalists; nobody remembers the news from the day before yesterday. A journalist is therefore more likely to be in error—but also to spot the unknown more quickly than a representative from intelligence or any other structured knowledge-producing activity.
- *Continuous interaction with an audience.* Journalism not only has a superior potential to sense what's new and separate potential scoops from the noise of trivial reporting. Journalists are also compelled to think in terms of their audience in a much more direct and concrete way than intelligence analysts. Dover and Goodman have observed a learning and adaptation process already under way in intelligence, yet intelligence still has some way to go, as illustrated by the Iraqi WMD case as well as more recent whistle-blowing.

Also, the mushrooming of content—perhaps better labeled "stuff"—on the web is challenging both traditional journalism and intelligence, and doing so in similar ways (an issue dealt with in more detail in the next chapter). What is striking, though, is that even quality media regard themselves as under existential threat from these new, mostly web-based services. By contrast, that recognition is still generally not apparent among intelligence services. A recent RAND project on collaborative tools in the intelligence community looked at the *New York Times* as a representing a profession kindred to intelligence.[88] The *New York Times* is not a cutting-edge social media practitioner. No wikis,[89] no micro-blogs, no fancy or modern in-house tool development are apparent. It encourages collaboration but relies on email. Yet all the people at the *Times*—the journalists, editors, and managers—stress how fundamentally journalism has changed in the past decade or so. Here are some of these ways, all of them suggestive for intelligence:

- *Collaboration* is the name of the game today. Most bylines are for multiple authors, often from different locations.
- *Speed* has forced the *Times* and its competitors to push to the edge of their comfort zone with regard to accuracy. The *Times* stressed that, perhaps more than its smaller and less-established competitors, it often had to give more weight to being right than being quick, and as a result often lost out to other media organizations in the race to publish first. Nevertheless, it, too, has seen an increase in its resort to the "correction" page.
- *Editing on the fly* is imperative in these circumstances.
- *Evaluation, career development, and incentives* have kept pace with the changing nature of the business. The paper takes pains to measure contribution, not just solo bylines, and it values a blog posting that attracts readership as much as a print article on the front page.
- *Changing impact measures* are changing value propositions. Appearing above the fold in the print version of the paper is still regarded as the place of honor among journalists; another enviable distinction is making the *Times* "10 most emailed" list.
- *Content/version control* are pervasive and central to the publication process. The ability to track versions, attribute to multiple authors, and control version aimed at print or e-publication is critically important.
- *Customer knowledge* perhaps most distinguishes the *Times* from the intelligence community, as Dover and Goodman suggest. It knows its readership exquisitely, to an extent most intelligence agencies can only dream about.

Weather Forecasting

When one of us (Treverton) was running the US National Intelligence Estimates (NIE) process, he took comfort from at least the urban legend that

---

[88] Gregory F. Treverton, *New Tools for Collaboration: The Experience of the U.S. Intelligence Community*, forthcoming from IBM Business of Government.

[89] Wikis are open-source online encyclopedias, which permit users to create and edit entries.

predictions of continuity beat any weather forecaster: if it was fine, predict fine weather until it rained, then predict rain until it turned fine. He mused, if those forecasters, replete with data, theory, and history, can't predict the weather, how can they expect us to predict a complicated human event like the collapse of the Soviet Union?[90] And when intelligence analysts evaluate a developing situation, contingent factors, and warning indications, they do resemble weather forecasters.[91] At least, the comparison "has been useful as a way to frame failure defined as inaccuracy in both fields."[92] In Marrin's words:

> In terms of emphasizing the relative significance of failure in each of the fields, Michael Herman has suggested that while "errors in weather forecasts do not call meteorology into question" sometimes errors in intelligence can and have called intelligence into question, and it is worth asking "whether there is anything special or different about intelligence failure" in the sense of not being able to forecast or predict the future accurately.[93]

Weather forecasters have at least one clear advantage: if even doctors aren't exempt from the element of human deception, weather forecasters are. Weather may surprise but doesn't deceive. Both weather forecasting and intelligence rely on models, but for weather the models are based on lots of data collected over many years. There is good reason to believe that the underlying models so constructed are stable over time; they will predict as well tomorrow as they explained yesterday. For this reason among many others, climate change is unsettling because it suggests that the parameters underlying current weather models are themselves changing. In contrast, the models used by intelligence analysts, often more implicitly than explicitly, are weaker ones derived from less history and fewer cases. The models can be useful in, say, predicting which countries will be instable but cannot predict when instability will come to a head in chaos or rebellion. There often is no confidence that the underlying parameters are stable; indeed, policy interventions are intended to make sure precisely that they aren't—something centuries of attempting to modify the weather haven't achieved. Finally, for some critical questions, such as understanding Al Qaeda in 2001, historical models seem of little use.[94]

[90] See Gregory F. Treverton, "What Should We Expect of Our Spies," *Prospect*, June 2011.
[91] Marrin, *Improving Intelligence Analysis*.
[92] Ibid., p. 103.
[93] Ibid.
[94] Ibid., p. 104.

Weather forecasters are suggestive for intelligence analysts in rejecting point predictions in favor of contingent estimates while attempting to communicate the level of uncertainty ("these cut-off lows are hard to predict"), and account for the effect of independent variables on dependent variables.[95] In conveying estimates, they are implicitly both humble and mindful of the limits of their listeners: chances of rain, for instance, are hardly ever predicted more finely than increments of 20 percent. Neither forecasters nor intelligence analysts routinely present postmortems, particularly if their forecast turned out to be wrong. Those postmortems might appeal only to the most analytic among us, which is probably why weather forecasters do not do them; but for those analytic types, learning what happened to invalidate the forecast would help in calibrating both the forecast and the forecaster next time around—another suggestion for intelligence.

Archaeology

Archaeology offers an intriguing analogy because it focuses on collection and its relation to analysis, while most of the other analogies speak more to analytic methods, though sometimes in the presence of fragmentary data or deception. It is a nice antidote to the general proposition that intelligence failures are analytic, a failure in mindset or in "connecting the dots," in that awful Jominian phrase. Yet "intelligence failures are indeed a problem of collection, because collectors are seldom able to produce the substantial quantities of relevant, reliable data necessary to reduce uncertainty."[96] Archaeology is most akin to intelligence as puzzle solving. For it, the puzzle is "what was that civilization (or tribe or village) like?" The information it has is inevitably fragmentary, literally the shards of implements or decoration or other objects that have survived the centuries, sometimes the millennia. Without some careful process for dealing with these remains, the archaeologist will be tempted to take creative leaps from shards to conclusions about the civilization, leaps that may be right but are likely to be wrong.

David Clark outlines that process as one of four steps:

1. What is the range of activity patterns and social and environmental processes that once existed—that is, what the archaeologist seeks to understand?
2. What sample and traces of these would have been deposited at the time?
3. What sample of that sample actually would have survived to be recovered?

---

[95] Marrin, "Best Analytic Practices from Non-Intelligence Sectors."

[96] See Matthew C. Pritchard and Michael S. Goodman, "Intelligence: The Loss of Innocence," *International Journal of Intelligence and CounterIntelligence* 22, no. 1 (2008): 147–164.

4. What sample of that sample actually was recovered?[97]

For intelligence, the process is a reminder of how much information that would be useful isn't likely to be there, and thus an antidote to the temptation to focus on the information it has, neglecting the information that's missing. In that regard, intelligence is both advantaged and handicapped by comparison to archaeology. Its main advantage is that it isn't dealing with groups or events that are centuries old; in puzzle solving, it usually is interested in what is happening *now*. Intelligence may also be clearer about the link between step 1 and step 2—that is, what would be the visible or hearable traces of the activities it seeks to understand? Its main disadvantage is that it may not have direct access to those traces, as the archaeologist typically does. Rather, intelligence may have only overheard discussion of those traces, or have information about them only second- or thirdhand. Intelligence-as-puzzle solving usually concentrates on seeking the missing puzzle pieces. Too seldom is the puzzle considered in reverse: "how did the picture break down into the pieces being considered? Did it happen all at once or in stages? And what happened to the missing pieces?"[98]

For intelligence-as-puzzle solving, the first step would be to describe clearly the complete range of target activity that intelligence seeks to understand. Suppose that was the current state of Iran's nuclear weapons program. Then, the second step would be to describe the traces that would be deposited at the time by that range of activity. Notice that mistakes are possible at this stage. In trying to understand Iraq's nuclear weapons program before the first Gulf war, in 1991, intelligence analysts missed lots of traces because the activity they expected to see was a formal, coordinated nuclear weapons program. When they didn't find this, they concluded that Iraq wouldn't have a nuclear weapon before the late 1990s.[99] Indeed, analysts had puzzled over why Iraq would buy uranium when it didn't have enrichment capacity (after the Israeli raid on its Osirak reactor in 1981), and later over why it would buy triggering devices for nuclear weapons when it was so far from having the weapons

In fact, it turned out that Iraq had been pursuing a plan not so different from the original American Manhattan Project, pursuing lots of possibilities, especially for enriching uranium—including one that long since had been dismissed by the nuclear powers as outdated—in the expectation that many would fall by the wayside.[100] Yet the United States was a long way from the Manhattan Project; it knew what worked best and didn't have to worry

[97] D. Clarke, "Archaeology: The Loss of Innocence," *Archaeology*, 47, no. 185 (1973): 16.

[98] Pritchard and Goodman, "Intelligence: The Loss of Innocence," p. 152.

[99] National Intelligence Council, "How the Intelligence Community Arrived at the Judgments in the October 2002 NIE on Iraq's WMD Programs," Washington, DC, 2004.

[100] See Rhodes, *The Making of the Atomic Bomb* (New York: Simon and Schuster, 1986), p. 25.

about what components it wouldn't be allowed to buy. As a result, what intelligence saw in Iraq looked incoherent, and analysts missed the traces indicating that Iraq "had been much closer to a weapon than virtually anyone expected."[101]

Step two suggests an inventory of what might have been available, while step three parses that inventory by what intelligence could reasonably have done. Intelligence's familiar sources, the "INTs,"—as in human intelligence, or HUMINT, signals intelligence or SIGINT—will all have their limitations. Critical activities may have been done inside facilities, limiting the traces for imagery, and even if done outside, satellites or other collectors may not have been in a position to collect them. Adversaries may work hard to shield their internal communications, limiting those traces in principle, and even if available, communications not intercepted at the time will not be available later. Spying, HUMINT, is less time sensitive, which is why, in contrast to so many discussions of intelligence that emphasize HUMINT as a way to get inside an opponent's head, it really is more valuable against capabilities than intentions. Intentions are often fleeting, and if a spy is not in the critical meeting or can't report on it to his handler, all is lost. By comparison, if a spy can't obtain a critical puzzle piece about capabilities today, she may be able to tomorrow, and the piece will still be of interest.

As a result, at step four, it may be possible to revisit traces that remain available at step three. Spies can be tasked to look for particular evidence or to recall things not deemed important originally, though memories will fade and self-interest will shape those that remain. So, too, if step three suggested an image that might be enduring or communications within the adversary that might continue, those too would become traces that might be revisited. For intelligence, though, the payoff probably lies mostly with steps two and three: if x had occurred (or was planned), what traces should we have seen? In that sense, the Iraq WMD fiasco and the failure to warn of the 1973 Egyptian attack across the Suez Canal are mirror images: in the first, when the traces of those WMD programs failed to appear, the most straightforward conclusion—the programs don't exist anymore—was disregarded in favor of more complicated explanations of deception or camouflage. By contrast, when in the runup to the 1973 war plenty of traces of impending attack did appear, the straightforward conclusion—an attack is imminent—was disregarded in favor of more complicated arguments about why Egypt wouldn't attack until it had the wherewithal to win.

---

[101] Iraq had pursued multiple uranium enrichment technologies, including a centrifuge program and the outdated electromagnetic isotope separation (EMIS) process. See National Intelligence Council, "How the Intelligence Community Arrived at the Judgments in the October 2002 NIE on Iraq's WMD Programs," p. 7.

Coda: Neuroscience

This chapter has scanned across other intellectual domains for comparisons that will be suggestive for intelligence. In one area of science, however, developments might directly affect how intelligence does its work by enhancing capacity. That area is neuroscience. Since at least the Cuban Missile Crisis and the recognition of the problem of "groupthink," systematic attempts have been made to incorporate the lessons of psychology and social psychology into each step of the intelligence process.[102] Scholars have identified psychological heuristics and biases that people, including analysts, use to assess their environment and to estimate the probability that certain events might occur—tools that often lead to incorrect estimates.[103] Scholars have also used insights from social psychology to design processes—such as red teaming, alternative analyses, and scenario building—in order to "break out" of groupthink, conventional wisdoms, or analytic mindsets. Theorists and practitioners of intelligence have shown a marked sensitivity to the insights of the cognitive sciences, even if the systems they design are, like all human institutions, still flawed.

However, recent advances in cognitive and social neuroscience have been slower to filter into these discussions of the psychology of intelligence.[104] A wide-spectrum search of the intelligence literature found no articles referencing, except in passing, contemporary neuroscience research. A search of the neuroscience and neuropsychology literatures yields virtually nothing about intelligence gathering, analysis, or consumption, or national security. One of the few to look at these issues is bioethicist Jonathan Moreno at the University of Pennsylvania.[105] He summarizes some of the developments in neuroscience being explored for national security purposes, and he proposes ethical guidelines for using advanced neuroscience and neuroscience

---

[102] Irving L. Janis, *Groupthink: Psychological Studies of Policy Decisions and Fiascoes* (Boston: Houghton Mifflin, 1972). The classic study applying the insights from psychology to intelligence is Richards J. Heuer, "Psychology of Intelligence Analysis" (Washington, DC: Center for the Study of Intelligence, Central Intelligence Agency, 1999).

[103] Rose McDermott, "Experimental Intelligence," *Intelligence and National Security* 26, no. 1 (2011): 82–98.

[104] One admirable effort to update Richards Heuer came to naught due to the CIA's internal sensitivities. Led by a psychologist, Richard Rees, the effort produced *Handbook of the Psychology of Analysis*. Cleared as unclassified, the CIA still decided not to publish the book—because "official" publication by the CIA was thought to sanction comments that were critical, albeit intended as constructive criticism. The result was thus the oddity of an unclassified book that is available only on the CIA's highly classified computer system.

[105] See Michael N. Tennison and Jonathan D. Moreno, "Neuroscience, Ethics, and National Security: The State of the Art," *PLoS Biology* 10, no. 3 (2012).

technologies by the national security establishment, and how neuroscientists can engage the national security establishment productively.

So far, the applications being researched apply more to operations than analysis. For instance, brain-computer interfaces (BCI) convert neural activity into signals for mechanisms, ranging from prosthetics to communication devices. One project, Augmented Cognition, pursued by the US Defense Advanced Research and Projects Agency (DARPA), sought to take neurological information from warfighters and use it to modify their equipment.[106] A "cognitive cockpit" would customize pilots' cockpits in real time based on monitoring their neural activity. It would select the least burdened sensory organ to receive communication, rank information needs, and reduce distractions. In another example, portable binoculars would convert subconscious neurological responses to danger into consciously available information. Another would use intracortical microstimulation (ICMS) to permit a neurologically controlled prosthetic to send tactical information back to the brain in near real time—thus creating a brain-computer-brain interface. Yet another would detect deficiencies in a warfighter's brain function, then use transcranial magnetic stimulation (TMS) to enhance or suppress particular brain functions. While prosthetics typically are thought of as aids to the afflicted, they could also be used on healthy people, with, for instance, neurological activity triggering devices to enhance strength or endurance.

Drugs to promote focus or avoid fatigue, like Dexedrine, are well known, even if their effect is still controversial. Research is looking into whether TMS can enhance neurological functions in healthy people, and other lines are asking whether in-helmet ultrasound transducers or transcranial direct current stimulation (TDCS) could do the same with more precision. Other applications seek not to enhance activity but to depress it. For instance, if memory of a traumatic event could be dampened, soldiers might be less likely to suffer from post-traumatic stress disorder (PTSD).

Another line of neuroscience research that has direct relevance to intelligence is dealing with deception. The shortcomings of the polygraph are well known, and several lines of research are seeking to overcome these. One, "brain fingerprinting," uses EEG (electroencephalography) to monitor a particular brain wave, the P300, a so-called event-related potential associated with the perception of a meaningful stimulus. The process is thought to hold the potential for confirming the presence of "concealed information." So, too, the hormone oxytocin is thought to enhance generosity and trust—obvious advantages for interrogators. Yet the long quest for a "truth

---

[106] This summary is from Ibid.

serum" warrants caution. As one study nicely put it: "the urban myth of the drugged detainee imparting pristine nuggets of intelligence is firmly rooted and hard to dispel."[107] The historical track record of abuse and gross violation of human rights in the wake of attempts to refine methods to extract information remains a moral downside in intelligence.

[107] J. H. Marks, "Interrogational Neuroimaging in Counterterrorism: A No-Brainer or a Human Rights Hazard," *American Journal of Law & Medicine* 33 (2007): 483–500.

6

Common Core Issues

Intelligence Analysis, Medicine, and Policy Analysis

This chapter seeks to shed light on the future of intelligence by looking at common issues that arise both for it and for the practice of medicine, as well as for policy analysis as it is being developed in social science departments and public policy schools of academia. The most visible common issue for both intelligence analysis and policy analysis is uncertainty. As major policy issues become bigger, messier, and more interconnected, uncertainty about policy approaches increases. That uncertainty is endemic to intelligence analysis. In that sense, other realms are becoming more like intelligence, or perhaps uncertainty is growing in all of them apace. What uncertainty, risk, and probability mean in the two areas need to be decoded. For the second core issue in common—what the military calls "blue," our own actions—the timing goes in the other direction. Almost by definition, if policy science is addressing a big issue, "we," those who will be affected by it, will be a major consideration in the analysis; in climate change, for instance, the United States is far and away the largest contributor of carbon per capita. For intelligence, however, the emphasis on blue is newer. Especially for the United States, "intelligence" meant "foreign intelligence" and stopped at the water's edge, as Chapter 3 discussed. Now, though, terrorists and other transnational threats are asymmetric, seeking "our" vulnerabilities, and thus those threats cannot be understood without analysis of us, turning intelligence assessment into "net assessment" very much akin to what policy science does for big policy issues. A third common core issue is new for both, though much more awkward for intelligence: the increasing transparency of a world where our cellphones locate us wherever we are and our contrails across the web reveal details about us down to our taste in good cigars or bad pictures. Policy science in many places has lived in a world of required open meetings, but the change in degree is notable when, for example, social media allow constituents to share views (and organize protests) instantaneously. For intelligence, by nature closed and passive, it is a sea change.

The Rise of Uncertainty

Because uncertainty always has been endemic to intelligence, it is both harder to discern the increasing complexity and, perhaps, easy to exaggerate it. In retrospect, the Soviet Union *was* pretty predictable; it surprised us most by collapsing. Yet it did not seem so slow moving at the time, a reminder that our perceptions may change more slowly than reality. After all, between the Soviet invasion of Afghanistan, perhaps the high water mark of "evil empire," and the end of that empire lay only a scant dozen years. And while the threat from transnational actors, especially terrorists, does seem less predictable and less bounded than the Soviet threat, analysts are beginning to provide some shape to the threat, especially as it becomes more homegrown, hence more like organized crime.

To be sure, puzzles remain important for intelligence: consider Osama bin Laden's whereabouts. And with time and experience, some of the complexities associated with, for instance, the terrorist target are being resolved toward mysteries. Still, the increase in complexity, hence uncertainty, is underscored by analogies with two other areas of science—medicine and policy analysis. The analogies between intelligence and medicine are striking. In both cases, experts—doctors or intelligence analysts—are trying to help decision makers choose a course of action. The decision makers for intelligence are policy officials, for medicine they are patients or policy officials. In both cases, the "consumers" of the intelligence may be experts in policy or the body politic or their own bodies, but they are not experts in the substance of the advice, still less in how to think about uncertainty. Both intelligence analysts and doctors are often in the position of trying to frame "mysteries." For medicine as for intelligence, the "output" may be the expert, not any written analysis. But consider two possible flu pandemics, three decades apart.

On February 4, 1976, a US Army recruit collapsed and died during a forced march at Fort Dix, New Jersey.[1] Scientists at the US Centers for Disease Control (CDC) determined that the virus was a form of the swine flu, the same virus believed to be the agent of the 1918–1919 global flu pandemic in which 500,000 Americans died. The CDC epidemiologists had been alert to the outbreak of some kind of flu pandemic, since then-current theories predicted that one would occur about once a decade, and the last one had been in 1968. They feared an outbreak would be disastrous since after 60 years the population's natural sources of resistance would be very low. The doctors

---

[1] This account is from Richard E. Neustadt and Harvey V. Fineberg, "The Swine Flu Affair: Decision-Making on a Slippery Disease," in *Kennedy School of Government Case C14–80–316*, Harvard University, Cambridge, MA.

found themselves in a position familiar to intelligence analysts, trying to convey their own subjective judgments about probability to policymakers who were uncomfortable with probabilities, especially low ones connected to very bad outcomes. In this case, the doctors erred on the side of professional caution and said simply that they couldn't quantify the probability. In fact, many of the doctors privately thought the probability was low. But what policymakers heard was different. One common description of the threat was "a very real possibility," and one policy official said: "The chances seemed to be one in two that swine flu would come."[2] So an unknown probability had become an even chance.

In the following weeks, no new cases of swine flu were reported, but investigations of "old" cases at Fort Dix revealed nine more cases of swine flu, bringing the total to 13. However, not only had the chances become heard as even but the analogy with 1918 was extended to suggest a million Americans might die.[3] By the time of the first major meeting with President Ford, March 22, even skeptics in the Office of Management and Budget were coming to the view that the president had little choice but to set in motion a major vaccination campaign. At that meeting, when the suggestion of a White House session with eminent medical experts came up, Ford embraced it eagerly, deferring his final decision. When the blue ribbon panel met two days later, the administration's real coup was the inclusion of both Jonas Salk and Albert Sabin, the two medical heroes of the fight against polio—who, however, had little love lost for each other. White House officials saw their presence as a test—if *they* agreed on something, it must be right. For others, though, the presence of the superstars suggested that the meeting was orchestrated, despite the president's repeated questioning whether anyone present disagreed. At the end of the meeting, flanked by Salk and Sabin, the president announced a program to vaccinate every American against swine flu.

Like many emergency programs, this one was dogged by problems from the start. It turned out that 200 million doses couldn't be produced by the fall, and field trials on children were disappointing. Still, more than 40 million Americans were inoculated by mid-December, twice the number inoculated for any other flu. A few elderly people died after receiving the shots, what are called "coincidental deaths" but are hardly good publicity. Finally, cases of a paralysis called Guillain-Barré began to emerge and be connected to the swine flu vaccine. What didn't emerge were more cases of swine flu. On December

[2] As quoted in "Swine Flu (E): Summary Case, Kennedy School of Government Case C14–83–527," Harvard University, Cambridge, MA, p. 8.

[3] In fact, while flus are viruses, many of the deaths are caused by respiratory infections, and antibiotics are available to treat those today but not in 1918.

16, the president reluctantly agreed to end the program. With the exception of the Fort Dix recruit, no one else had died of swine flu.[4]

Whether swine flu would become a pandemic was uncertain, but that uncertainty was relatively straightforward. In retrospect, a sensible strategy might have hedged, waiting to improve the probability estimates, perhaps stockpiling vaccine but only beginning vaccinations when signs of spread appeared. Had it become a pandemic, it would have become a global concern, but it was relatively bounded. "Jet spread" of disease was possible—and even the 1918 flu had become worldwide—but was not yet ubiquitous and nearly instantaneous. The severe acute respiratory syndrome (SARS), another virus, shows the contrast 30 years later, in complexity and global spread of both disease and possible responses.

In the near-pandemic that occurred between November 2002 and July 2003, there were 8,096 known infected cases and 774 confirmed human deaths. This resulted in an overall case-to-fatality rate of 9.6 percent, which leaped to 50 percent for those over 65. By comparison, the case-fatality rate for influenza is usually less than 1 percent and primarily among the elderly, but it can rise many-fold in locally severe epidemics of new strains. The 2009 H1N1 virus, which killed about 18,000 people worldwide, had a case-fatality rate no more than.03 percent in the richer countries. [5]

SARS spread from Guangdong province in southern China, and within a matter of weeks in 2002 and early 2003 had reached 37 countries around the world. On April 16, the UN World Health Organization (WHO) issued a press release stating that a coronavirus identified by a number of laboratories was the official cause of SARS; the virus probably had originated with bats and spread to humans either directly or through animals held in Chinese markets. Once the virus was identified, every health professional became, in effect, an intelligence collector on the disease. WHO set up a network dealing with SARS, consisting of a secure website to study chest x-rays and to conduct teleconferences.

The first clue of the outbreak seems to have appeared on November 27, 2002, when a Canadian health intelligence network, part of the WHO Global Outbreak and Alert Response Network (GOARN), picked up reports of a "flu outbreak" in China through Internet media monitoring and analysis and sent them to the WHO. WHO requested information from Chinese authorities on

---

[4] It turned out that when the recruit collapsed, his sergeant had given him mouth-to-mouth resuscitation without contracting swine flu!

[5] The fatality number is from the World Health Organization; the case fatality numbers from studies in Britain and the United States. The WHO acknowledges that many more people may have died from the flu, mostly in Africa, isolated from both treatment and accounting.

December 5 and 11. It was not until early April that SARS began to receive much greater prominence in the official media, perhaps as the result of the death of an American who had apparently contracted the disease in China in February, began showing symptoms on the flight to Singapore, and died when the plane diverted to Hanoi. In April, however, accusations emerged that China had undercounted cases in Beijing military hospitals, and, under intense pressure, China allowed international officials to investigate the situation there, which revealed the problems of an aging health care system, including increasing decentralization, red tape, and weak communication. WHO issued a global alert on March 12, followed by one from the US Centers for Disease Control and Prevention.

Singapore and Hong Kong closed schools, and a number of countries instituted quarantine to control the disease. Over 1,200 people were under quarantine in Hong Kong, while in Singapore and Taiwan, 977 and 1147 were quarantined, respectively. Canada also put thousands of people under quarantine. In late March, WHO recommended screening airline passengers for SARS symptoms. Singapore took perhaps the most extreme measures, first designating a single hospital for all confirmed and probable cases of the disease, then requiring hospital staff members to submit personal temperature checks twice a day. Visiting at the hospital was restricted, and a phone line was dedicated to report SARS cases. In late March, Singapore invoked its Infectious Diseases Act, allowing for a 10-day mandatory home quarantine to be imposed on all who might have come in contact with SARS patients. Discharged SARS patients were under 21 days of home quarantine, with telephone surveillance requiring them to answer the phone when randomly called.

On April 23, WHO advised against all but essential travel to Toronto, noting that a small number of persons from Toronto appeared to have "exported" SARS to other parts of the world. Toronto public health officials noted that only one of the supposedly exported cases had been diagnosed as SARS and that new SARS cases in Toronto were originating only in hospitals. Nevertheless, the WHO advisory was immediately followed by similar advisories from several governments to their citizens, and Toronto suffered losses of tourism. Also on April 23, Singapore instituted thermal imaging screens on all passengers departing from its airport and also stepped up screening at points of entry from Malaysia. Taiwan's international airport also installed SARS checkpoints with an infrared screening system similar to the one in Singapore. The last reported SARS case in humans was June 2003, though the virus may remain in its animal hosts.

It took more than three months from first information about the disease to a global alert. It was then another month until the virus was clearly identified. The time delay may have had something to do with China's dissembling about the extent of the disease, but it also demonstrates that the cause of any outbreak—whether natural or terrorist—may take some time to identify. Once the virus was identified, however, virtually every health care professional in the world became a potential collector of intelligence on the disease, illustrating the other side of the complexity coin; this is suggestive for intelligence as well. The global network was a form of "crowd-sourcing," though the label was not used at the time. The worldwide web permits people or organizations—in principle including intelligence agencies—to enlist the help of strangers in solving problems, an issue this chapter turns to in the conclusion. Crowd-sourcing requires openness, which is the antithesis of intelligence. It also requires incentives. In the case of SARS, the public-spiritedness of health professionals was sufficient to induce them to participate, but that is not the case for other attempts at crowd-sourcing.

Policy Analysis and Intelligence Analysis

Policy science offers another parallel to growing uncertainty and complexity in intelligence. The traditional paradigm of policy analysis might be caricatured as "*the* study for *the* decision."[6] To continue the caricature, this traditional paradigm presumed there was an authoritative person or group making the decision, and that the person or group had well-defined objectives. There were relatively well-defined alternatives of policy, or treatment or intervention. The uncertainty arose primarily in linking those alternatives to their effects. In an effort to reduce that uncertainty, the decision maker turned to an analyst, much as foreign policy officials sometimes turn to intelligence analysts to try to reduce the uncertainty they face. The main difference is that the traditional policy analysis paradigm presumed that the decision maker was specifically interested in analysis of given alternatives, while intelligence often is in the position of trying to reduce uncertainty without being privy to alternatives in the minds of foreign policy decision makers. More formally, the decision maker in traditional policy analysis is asking, In light of my objectives, what are likely costs and benefits of each alternative? A more quantitative decision maker might put the question in terms of a probability distribution of costs and benefits.

With the question in hand, the analysts first assemble the variables that affect benefits and costs, usually employing theories about programs that are widely accepted and understood. Those analysts collect complete and reliable

---

[6] This phrase and much of the following owes to Robert Klitgaard, "Policy Analysis and Evaluation 2.0," unpublished paper, 2012.

data, then employ established analytic techniques—drawing on economics, statistics, experimental methods, and mathematical modeling—to assess the costs and benefits of each alternative. More sophisticated analysis might take into account variables internal to the process, such as bureaucratic politics, the political process, time constraints, and the assets and vulnerabilities of the decision maker's position. The result will be a recommendation, one presented in a well-crafted report, perhaps with a Power Point briefing and decision memo. In this caricature, the decision maker will digest the analysis, then make a decision. If the analysis and decision are successful, the results will be consistent with the predictions of the analysis. The policy analysis will be useful and used. In another parallel with intelligence, however, at the end of the process no one is likely to ask, Was it worth it? Was the decision advantage provided by the analysis worth its cost? In intelligence, the question doesn't get asked primarily because intelligence is a free good to policy makers. In policy analysis, the presumption is usually that analysis is cheap by comparison to the alternatives being assessed.

Policy analysis, in this caricature, has one important advantage over most intelligence, one referred to earlier: it is operating on well-defined alternatives. Conversely, for intelligence too much of the time the question at issue is messy: what's going to happen? Yet even with that advantage, policy analysis still confronts formidable challenges, ones with visible parallels for intelligence. As Nobel Prize winner Robert Solow put it: "Measuring the benefits of some social policy intervention rests fundamentally on our ability to disentangle the true effects of the intervention from all the other things that are going on at the same time."[7] And that is a formidable task indeed.

Scientific studies—in medicine, for instance—seek internal validity; thus the study population should be as homogenous as possible. By contrast, policy analysis wants external validity, seeking approaches that will work across time and space. But clinical trials in medicine may not work in everyday situations, particular with populations very different from those in the trial. So too the alternative or treatment may look the same but be different in different places and times. What works well in a rich-country university medical center may not in a poor-country community hospital. This fact can be a valuable source of adaptation, but those adaptations will, at best, undercut the original policy analysis and could, at worst, open the way for adjustments that are not benevolent and might even be corrupt. So too, while the paradigm assumes that objectives are both easily agreed and measured, that is not the case, and never was. More on that later. Nor can models and multivariate analysis come to the

---

[7] R. M. Solow, "Forty Years of Social Policy and Policy Research," Inaugural Robert Solow Lecture (Washington, DC: Urban Institute, 2008), p. 4.

rescue. Most important—and perhaps most relevant to intelligence—almost all models base their assessments of future policy change on the past. Yet complexities don't behave in accord with past patterns, or at least not visibly so, practically by definition.

These challenges, for both intelligence and policy science, are compounded the farther "upstream" the decisions to be enlightened by analysis are contemplated.[8] Policy science dealing with the environment, for instance, often confronts choices about interventions quite far upstream—that is, decades or more before the serious effects of climate change are forecast to occur—and these choices must be made lest inaction foreclose options. There is, for example, so much momentum to the buildup of carbon in the earth's atmosphere that whatever effect that buildup has on climate change will continue for decades even if carbon emissions were dramatically reduced tomorrow. Yet for both the environment and intelligence, early interventions are, almost by definition, characterized by limited information. Equally as important, the earlier the assessment, the more contingent it must be. A long trail of mysteries, most of them based on human behavior, will intervene between the assessment and the event far in the future.

Finally, perhaps the most obvious way in which the traditional paradigm of policy analysis is a caricature is also one very parallel to intelligence: there is little evidence that policy analysis is decisive, or even heeded, including when it is right. Robert Klitgaard tells a charming story about bearding the US deputy secretary of defense on the phone during a class at the RAND Graduate School when he was dean. The secretary praised RAND's landmark study of gays in the military in 1993. Yet the Pentagon had not taken its advice (let gays serve), instead opting for "don't ask, don't tell." Why the praise? In the secretary's words: "That study is the gold standard on this issue. Every time the issue comes up, which is often, and whenever new folks come along, like new senators or congressmen, and they ask about this issue, we give them the RAND report. It lays out the alternatives and the pros and cons. It gives everyone an authoritative, unbiased look at the issue. What more could we ask of a policy study we commissioned?"[9] For both policy analysis and intelligence, going unheeded doesn't necessarily mean going unheard.

If the past of policy analysis, as least as it is remembered, introduced uncertainty mostly in the link between action and result, the future of policy science recognizes uncertainty and complexity at virtually every stage. In these respects it more and more resembles intelligence. If the problem used to be perceived as

---

[8] See Mark Phythian, "Policing Uncertainty: Intelligence, Security and Risk," *Intelligence and National Security* 27, no. 2 (2012): 198 ff.

[9] Quoted in Klitgaard, "Policy Analysis and Evaluation 2.0."

well defined, now it is not; often it's poorly understood. If objectives and alternatives used to be taken as givens, now the former are murky and the latter perhaps incomplete. If relevant data were available and valid, and the process of data generation was understood, none of those is any longer the case. If the context was reduced to a standard, now it is highly variable and perhaps complex. If the decision maker was unitary, now it is multiple, usually with no single or decisive "decisions." If the relationship between policy analysts and policymakers was arm's length, emphasizing objectivity, now it emphasizes problem clarification and learning in both directions. Turn first to the implications of "us." Part of this cluster is newer for intelligence than for policy science—the greater effect of "our" actions on the problem. The other part is new for both—the greater number of stakeholders, if not decision makers, and the changed relations between the two.

The Importance of "Us"

Traditional policy analysis, what Klitgaard calls "policy analysis 1.0," differed from intelligence analysis in that it had to take into account "us," virtually by definition. After all, if the point of the analysis was to make good choices about people's welfare, those people, the "us," could hardly be excluded. Yet even policy analysis 1.0 confronted a challenge on that score, one akin to intelligence. It had held that if assessing effects of proposed alternatives depends on models based on past behavior, that analysis is suspect if there is reason to believe the behavior being modeled has changed. In actuality, the challenge is still harder: the policy change or intervention itself will also change other variables in the model. This is what experimental psychology calls the "Hawthorne effect" and economics the "Lucas critique."[10] Analysis of effects is frustrated because the policy initiative itself will set in motion other changes, usually unanticipated, thus invalidating predictions about the policy change based on an unchanging model.

---

[10] In the classic studies at the Hawthorne Works of Western Electric in 1924–1932, experiments sought to see if changes in the workplace, such as more or less light, would lead to more worker productivity. What ensued was that almost any change improved productivity but only during the course of the experiment. The conclusion was that the attention of the initiative itself had motivated workers. See Henry A. Landsberger, *Hawthorne Revisited: Management and the Worker, Its Critics, and Developments in Human Relations in Industry* (Ithaca: New York State School of Industrial and Labor Relations, 1958). For economics, the classic study is R.E. Lucas, Jr., "Econometric Policy Evaluation: A Critique," in K. Brunner and A.H. Meltzer, eds. *The Phillips Curve and Labor Markets, Carnegie-Rochester C Carnegie-Rochester Conference on Public Policy*, Vol. 1 (Amsterdam: North Holland), pp. 19–46.

As discussed in Chapter 3, traditional intelligence analysis, especially in the United States, jumped over this challenge by taking "us" out of the analysis. This first dimension of "us"—the effect of our actions on the question under analysis—has led to first steps in a number of countries. If intelligence on many issues has become net assessment—how does the intersection of foreign circumstances and our actions, real or potential, net out?—those first steps for many countries have been to create threat assessment centers.[11] Those mostly are all-hazard, trying to prepare both for nature's challenges and manmade threats. Analysts in these threat assessment centers have developed methods to try to assess threats in light of the society's vulnerability. Most of them are all-government, bringing in "domestic" as well as "foreign" agencies. To that pattern the US National Counterterrorism Center (NCTC) is an exception; given its origins in foreign intelligence, it still retains a foreign focus despite participation by the Federal Bureau of Investigation (FBI). Indeed, in most of the countries, again with the United States as an exception, the threat assessment operation is part of or is reliant on the domestic intelligence agency. For its part, intelligence in the United States has become less timid in assessing the implications of alternative US actions—provided it knows what the alternatives are. It is also a little less timid about what is called "opportunity analysis," looking for ways in which US action might make a positive difference given US interests. There is a long way to go.

The other dimension of "us" is new enough for policy analysis but very new for intelligence. The similarities and differences as they seek to adapt are suggestive. Intelligence, especially at the national level, used to have a relatively small number of consumers, most of them politico-military officials in the national government. Now, in principle, "consumers" extend to police on the beat and to private sector managers of "public" infrastructure. In the United States, police are divided into 18,000 jurisdictions, and total more than 700,000 sworn officers. In the United States, the approach to dealing with that vastly expanded set of consumers is labeled "information sharing." That is about the worst way to conceive of the challenge. It implies that the problem is technical, building bigger information pipes, and that agencies "own" their information; it is theirs to share as they see fit. It also implies that the sharing is in one direction, downward from the federal agencies to state and local authorities, and private citizens. Yet the fundamental challenge is to reshape how intelligence services, and the government in general, think of information, and how it should be produced, used, and controlled.

---

[11] For a survey, now several years old, see Gregory F. Treverton and Lisa Klautzer, *Frameworks for Domestic Intelligence and Threat Assessment: International Comparisons* (Stockholm: Center for Asymmetric Threat Studies, Swedish National Defence College, 2010).

The newer innovation in the United States is "fusion centers." Similar to the older FBI-sponsored Joint Terrorism Task Forces (JTTFs), fusion centers co-locate analysts from several agencies to facilitate the integration of several streams of information. They are meant to fuse foreign intelligence with domestic information to facilitate improved decision making on issues of counterterrorism, crime, and emergency response.[12] Their membership is determined by local and regional need and security priorities. Although they are state-created and state-based law enforcement entities, most fusion centers are partially funded by the US Department of Homeland Security, which has also seconded intelligence analysts to them. They numbered about 70 at the end of 2012.

The fusion centers are intended to complement the FBI-sponsored Joint Terrorism Task Forces, which also bring together state and local authorities with federal officials. If JTTFs work on cases once identified, the fusion centers are meant to assemble *strategic intelligence* at the regional level and pass the appropriate information on to the investigators in the task forces. In practice, the fusion centers are very much a work in progress, displaying all the challenges of jurisdiction. Their missions vary based on regional requirements and resources. Structurally, they differ considerably from one another in organization, management, personnel, and participation. Communication between centers ranges from problematic to nonexistent. The centers' access to technology and intelligence information is uneven. Not all fusion centers even have statewide intelligence systems. According to the single congressional review, now slightly dated but still basically accurate: "The flow of information from the private sector to fusion centers is largely sporadic, event driven, and manually facilitated."[13] They also do not all have access to law enforcement data. The problem of diverse platforms that was widely criticized immediately after 9/11 still exists.

Security clearances remain a bar to sharing, though the situation is improving. In 2007, roughly half the staff of the average fusion center (totaling about 27), had Secret clearances, about 6 had Top Secret and 1 had Secret Compartmented Intelligence (SCI).[14] As for the JTTFs before them, the obstacles the fusion centers encounter in getting local officers cleared can be as much personal as procedural: more than one crusty local officer has said to a young FBI agent, with more or less scorn in his voice, *"you're* going to clear

---

[12] This and the next paragraph draw on Todd Masse, Siobhan O'Neil, and John Rollins, *Fusion Centers: Issues and Options for Congress* (Washington, DC: Congressional Research Service, 2007).

[13] Ibid., p. 29.

[14] Ibid., p. 26.

me?"[15] Because of the huge number of information systems and the resulting duplication, analysts are inundated with floods of information of variable
quality. A review of the Los Angeles fusion center, called the Joint Regional
Intelligence Center, said this: "an overbroad intelligence report distributed by
the LA JRIC . . . offers no perceived value for police agencies and is not frequently used to deploy police resources. Typically, the LA JRIC collects and
distributes open source material or newsworthy articles in an effort to inform
their clients. As a result local independent chiefs feel they cannot use the intelligence to increase operational capacity or deploy police resources to combat
crime or terrorism."[16]

Further, although much rhetoric is expended on the importance of a cycli-
cal information flow between the fusion centers and the federal intelligence
community, the information cycle tends to be rather one directional. From the
perspective of the fusion centers themselves, the value of information becomes
opaque once it is sent to the federal intelligence agencies.[17] This lack of a feedback loop creates both resentment and inefficiency in the relationship between
federal entities and the fusion centers. Things have improved from the days
when local authorities reported that if they called the FBI, they never received
a call back, but there is a long way to go. In the counterterrorism realm, in particular, it is not clear in many jurisdictions how or if tips from state and local
authorities that are not deemed to rise to the level of JTTF cases get registered
in the system.

Moreover, there has been lingering concern about issues of civil liberties
and fusion centers, mainly because of the public perception that they lack
transparency in their operations. A range of reports, including one published
by the American Civil Liberties Union (ACLU) in December 2007, brought
up concerns that the expansion of intelligence gathering and sharing in these
centers threatens the privacy of American citizens.[18] Some of the concerns
raised by the ACLU stem from the wide range of participants involved in the
fusion centers, including the private sector and the military. The report argues
that breaking down the many barriers between the public and private sectors,

[15] See Daniel M. Stewart and Robert G. Morris, "A New Era of Policing? An Examination of
Texas Police Chiefs' Perceptions of Homeland Security," *Criminal Justice Policy Review* 20, no. 3
(2009): 290–309.

[16] Phillip L. Sanchez, *Increasing Information Sharing among Independent Police Departments*
(Monterrey, CA: Naval Postgraduate School, 2009), pp. 3–4.

[17] This discussion draws on Henry H. Willis, Genevieve Lester, and Gregory F. Treverton,
"Information Sharing for Infrastructure Risk Management: Barriers and Solutions," *Intelligence
and National Security* 24, no. 3 (2009): 339–365.

[18] See Michael German and Jay Stanley, *What's Wrong with Fusion Centers?* (New York:
American Civil Liberties Union, 2007).

intelligence and law enforcement, and military and civil institutions could lead to abuses of private and other civil liberties, particularly if a strong and clear legal framework isn't established to guide operations within the centers.

The fusion centers are likely to take a variety of paths in the future. Virtually all are moving away from a singular focus on terrorism to an all-crimes, or even all-hazards, approach—probably a happy spillover from the preoccupation with terrorism to broader policing. In some places, like the state of Iowa, where terrorism is a minor threat, the center explicitly is in the business of driving intelligence-led policing. Other centers will simply fade away if and as federal support diminishes and contributing public safety agencies decide their personnel are more valuable "at home" rather than seconded to a fusion center. A 2012 congressional assessment was too narrow—limiting itself to the centers' contribution to the nation's counterterrorism effort—but still its criticism was not off the mark. It found "that the fusion centers often produced irrelevant, useless or inappropriate intelligence reporting to DHS [Department of Homeland Security], and many produced no intelligence reporting whatsoever."[19] Looking at a year's worth of reports, nearly 700, the investigation "could identify no reporting which uncovered a terrorist threat, nor could it identify a contribution such fusion center reporting made to disrupt an active terrorist plot."

To stretch thinking about dealing with a mushrooming "us," Table 6.1 compares sharing arrangements in three different policy realms—terrorism, infectious disease, and natural disasters.

For intelligence, the difference in "us"—be they consumers or participants—is qualitative; categories of people that would never have been thought of as users of intelligence now need it. For policy analysis 2.0, in Klitgaard's phrase, the change is less stark but still notable: more stakeholders care about a set of decisions and more may be organized to watch the decision process and to try to influence it. For policy analysis, the expanded numbers merge with the third common core concern, transparency. If it always was a fiction that an authoritative decision maker could make a choice based on *the* study, it is more and more of a fiction. Processes for reaching decisions need to be more and more participatory and transparent even at the cost of ponderousness and delay. In public sector decision making, process is as important as outcome, perhaps more so.

Start with a lament common to both intelligence and policy analysis: decision makers don't heed us. Conclusions about policy analysis abound that, with a few changed words, could come from intelligence practitioners:

[19] Senate Committee on Homeland Security and Governmental Affairs, Report of the Permanent Subcommittee on Investigations, *Federal Support for and Involvement in State and Local Fusion Centers* (Washington, DC, 2012), p. 2.

Table 6.1 **Sharing in Different Issue Realms**

Culture	System	Technical Expertise and Knowledge	Data Ownership and Management
Terrorism	Closed	Embedded within organizations	Security-limited at national level but more creation and use at local/state levels
Infectious Diseases	Primarily open (data secured to protect privacy but more "securitization")	Dispersed across governmental and NGOs (research primarily federally funded)	Tiered data ownership at many levels; analyzed data are public
Natural Hazards	Open	Same as above	Data disseminated across agencies; analyzed data broadcast to public

Source: http://www.stimson.org/pub.cfm?ID = 730

Even a cursory review of the fate of social science research, including policy research on government-defined issues, suggests that these kinds of expectations [of direct and immediate applicability to decision making] are wildly optimistic. Occasional studies have direct effect on decisions, but usually at the relatively low-level, narrow-gauge decisions. Most studies appear to come and go without leaving any discernible mark on the direction or substance of policy.[20]

If the laments are similar, so are the frustrations in trying to learn lessons in order to do better. One experiment in policy analysis took advice from the famous statistician Frederick Mosteller: "People can never agree on what benefits and costs are. But they can and do agree on specific examples of outrageous success and outrageous failure. Find these among your projects. Study them. Compare them. Share your results, and learn some more."[21]

[20] C. H. Weiss, "The Many Meanings of Research Utilization," *Public Administration Review* 39, no. 5 (1979): 426.
[21] As quoted in Klitgaard, "Policy Analysis and Evaluation 2.0," p. 18.

The US RAND Corporation made one effort to apply Mosteller's dictum by looking for policy analyses that made a difference—programs that were agreed to be successes if not necessarily outrageously so. The effort began with examples from RAND's own work, then broadened to include programs of policy research institutes around the world. The list of fifteen successes ranged widely across issue areas and countries. Of the fifteen, however, only one approximated the policy analysis 1.0 caricature of "*the* study leading to *the* decision." What the others did provides a list that resonates with intelligence, especially as it seeks to reshape itself in dealing with a larger group of users: raised a hidden issue into prominence; shifted the rhetoric about an issue, which in turn helped people see reality in new ways and be more creative in thinking about policy objectives and alternatives; helped people see the objectives more completely; helped people see the set of alternatives more clearly, including some they had not envisioned; "complexified" an oversimplified debate; summarized an array of results and clarified the major categories of users or outcomes; provided a new, factual foundation for debate by introducing data that limited "spin" and helped people focus on reality; engaged multiple decision makers in ways that fostered collaboration.[22]

Also intriguing for the new world of intelligence were the conclusions from this initiative that might be called interactive or interpersonal. To quote Klitgaard:

> In many of the high-impact projects, strong *personal relationships* between researchers and policymakers made a big difference, ranging from identifying better questions to making sure the research arrived at the right time in the right ways. Analysts built trust with decisionmakers, leading to two outcomes. First, the decisionmakers had more confidence in the quality and objectivity of the analysts' work. Second, the decisionmakers felt free to pose more sensitive but also more relevant questions to the analysts. The result: research that was more useful and "used." How to develop the right kinds of relationships and still retain independence and objectivity is a challenge for the craft of applied research.
>
> Especially with complex problems featuring multiple stakeholders, a key role for high-impact policy research is what I call *convenings*. Stakeholders are convened, and the policy research catalyzes the stakeholders' creative, joint problem solving. The results can be deeper understanding of the issues and institutions, better decisions, stronger partnerships, and more realistic and therefore more successful

---

[22] Ibid., pp. 21–22.

implementation. All of these may go beyond the recommendations that any evaluator or policy analyst could have produced.[23]

Klitgaard's footnote at the end of his first paragraph about relationships and independence is also worth spelling out. It comes from Michael Quinn Patton:

> The importance of the personal factor in explaining and predicting evaluation use . . . directs us to attend to specific people who understand, value and care about evaluation, and further directs us to attend to their interests. This is the primary lesson the profession has learned about enhancing use, and it is wisdom now widely acknowledged by practicing evaluators. [24]

Indeed, Klitgaard's summary of Patton's distinction between traditional policy analysis and version 2.0 is provocative for intelligence (see Table 6.2).

For intelligence, not only has the set of possible collaborators and consumers exploded in number, but the nature of the threat and the multiplying uncertainty require a different kind of relationship to them. It also requires intelligence to rethink its "products." Neither "consumer" nor "customer" connotes the right sort of relationship between intelligence and the policy officials it seeks to serve. Of the two, consumer is better because intelligence is usually a free good, so its customers don't pay for it. Yet both connote a transaction, not a relationship. They suggest arm's length exchange ("over the transom" in a locution only those over 50 understand). Recall Pettee from Chapter 3: "The failure to define clearly the special province of 'strategic' or 'national policy' intelligence . . . meant in the past that the conduct of the work lacked all the attributes which only a clear sense of purpose can give." In some ways, perhaps, the Cold War was enough like hot war to justify both Kent's lumping together of intelligence in war and peace and the arm's length relationship between intelligence and policy. The major adversary was clear and clearly visible; most of the issues of interest were defined, as was the relationship of the United States to them; and strategic surprise was unlikely.

Yet none of that is true for the future of intelligence. Adversaries, particularly transnational ones, are a changing set. The issues are constantly shifting and often unclear. Uncertainty is rife. Surprise is likely. In this circumstance, policy officials cannot be consumers to be satisfied with products. They need

---

[23] Ibid., p. 23.

[24] M. Q. Patton, "Use as a Criterion of Quality in Evaluation," in *Visions of Quality: How Evaluators Define, Understand and Represent Program Quality: Advances in Program Evaluation,* ed. A. Benson, C. Lloyd, and D. M. Hinn (Kidlington, UK: Elsevier Science, 2001), p. 163.

Table 6.2 **Policy Analysis 1.0 versus 2.0**

Dimension	Policy Analysis 1.0	Policy Analysis 2.0
Situation when appropriate	Stable setting; causes of problems known and bounded; alternatives well understood; key variables are controllable, measurable, and predictable.	Complex, dynamic setting; not all causes known or bounded; not all alternatives well understood; multiple pathways possible; need for innovation and social experimentation.
Focus of evaluation	Top-down (theory driven) or bottom-up (participatory).	"Evaluation helps innovators navigate the *muddled middle* where top-down and bottom-up forces intersect and often collide."
Stance of evaluator	Independent; credibility depends on this.	Part of innovation team; facilitator and coach. Credibility depends on a mutually respectful relationship.
Valued attributes of evaluator	Methodological competence and rigor; analytical and critical thinking; external credibility.	Methodological "flexibility, eclecticism, and adaptability"; creativity; tolerance for ambiguity; teamwork and people skills.
Communication and facilitation	Written reports and briefings.	"Able to facilitate rigorous evidence-based reflection to inform action."

Source: Klitgaard, "Policy Analysis and Evaluation 2.0," p. 27. Based on Patton, "Use as a Criterion of Quality in Evaluation," in A. Benson, C. Lloyd, and D. M. Hinn, eds. *Visions of Quality: How Evaluators Define, Understand and Represent Program Quality: Advances in Program Evaluation* (Kidlington, UK: Elsevier Science, 2001), pp. 23–26. Quoted passages are from Patton.

to be partners in Kendall's "big job—the carving out of the United States destiny in the world as a whole."

The Imperative of Transparency

Social networking media, like Facebook and Twitter, are both a cause and a metaphor for the transparency that is rolling across both intelligence and policy science. Again, the change is one of degree for policy science, but it is virtually qualitative for intelligence. If the "us" has always been central to policy, so, too, advancing technologies over the centuries have expanded the number of people who could be engaged: notice the sweep from Gutenberg's press in the 1400s to the cassette recordings of Ayatollah Khomeini that were such a part of the Iranian revolution in the 1970s. In that sense, social media are only the latest in a long series of new technologies, and whether they will make a qualitative change in the ease of organizing for policy changes remains to be seen. It is clearer, though, that social media symbolize a wave that will transform the world in which intelligence operates.

Social media, part of the larger development of the worldwide web known as Web 2.0, are web-based services that facilitate the formation of communities of people with mutual or complementary interests, and provide them a means of sharing information with each other. They can put together people who do not know each other, and thus can promote the formation of communities of interest. In fact, the category of "social media" includes a wide range depending on how interactive they are and how much they are designed to connected people who may not know each other. For instance, email is interactive but normally connects people who already know each other. By contrast, websites certainly can connect people who don't know each other but are not typically very interactive. Blogs vary in how interactive they are and how open to new and unknown participants. Even today's hottest social media, Twitter and Facebook, are very different. The former is entirely open to new connections and, in principle, widely interactive; sheer volume, however, may limit how interactive it is in fact. By contrast, Facebook is designed primarily to promote interactions among people who already have some connection to each other, however scant.

The revolutions in devices and media amount to a sea change. A few statistics will drive home just how fast the information world is changing. Personal (non-work) consumption of information by Americans grew at 2.6 percent per year from 1980 to 2008, from 7.4 to 11.8 hours per day per person, leaving only 1.2 hours per day per person not spent consuming information, working (and also increasingly consuming information), or sleeping.[25] To be sure, traditional media—radio and TV—still dominate US consumption, accounting for about 60 percent of the daily hours people spend engaged. Yet computers and other ways of accessing the Internet flood around us.[26] Twitter grew by

[25] These and other statistics in this paragraph are from Roger E. Bohn and James E. Short, *How Much Information? 2009: Report on American Consumers* (San Diego: Global Information Industry Center, University of California, 2009).

[26] While much of this volume is explained by the use of graphics, the fact that it is interactive is consistent with the growing interest in and use of social media.

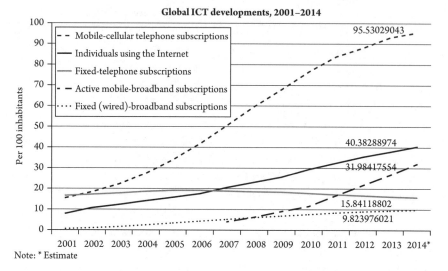

Figure 6.1 Mobile Phone by Comparison to Fixed Lines, Worldwide, 2001–2014.

1,500 percent in the three years before 2010, and it now has more than almost a quarter billion active users. Facebook reached 500 million users in mid-2010, a scant six years after being created in a Harvard dorm room.

Mobile, smartphones will be the norm for communications and sharing. As Figure 6.1 suggests, mobile phones already overwhelm traditional fixed phone lines, having grown in a decade from half a billion to almost five billion users.

The world's population was estimated in mid-2009 to be 6.8 billion, 27 percent of whom were under the age of 14. Thus, on average, there was one mobile phone for everyone over the age of 14. Of course, mobile phone ownership is not evenly distributed worldwide, but the trend toward ubiquitous ownership is obvious.

Even more portentous than the explosive growth in mobile phones is the revolution in their capabilities. The iPhone and other "smart" handsets let users gain access to the Internet and download mobile applications, including games, social networking programs, productivity tools, and more. These same devices also allow users to upload information to network-based services, for the purposes of communication and sharing. Smartphones, representing a manifold increase in processing power from ordinary mobile phones ("cell" phones or "feature" phones), accounted for over a third of the handsets shipped in North American in 2009, and some analysts estimate that by 2015 almost all handsets will be smart.[27] Mobile operators have started building

[27] Here, too, precise language is elusive. "Smartphone" seems the phrase of choice but is pretty tepid given the capabilities of the devices. PDA, or personal digital assistant, already rings out of date, while "hand-held computer" sounds clunky. On the revolution, see "The Apparatgeist Calls," *The Economist*, December 30, 2009.

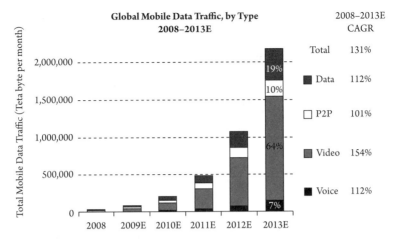

Figure 6.2 Video Driving Mobile Internet.

networks that will allow for faster connection speeds for an even wider variety of applications and services.

The other striking feature of social media's future is that it will be dominated by images, not words. That is plain with Facebook but is increasingly the case for Twitter as well. As Figure 6.2 indicates, video is driving the growth of the mobile Internet.

Reading, which had been in decline due to the growth of television, tripled from 1980 to 2008, because it was overwhelmingly the preferred, indeed necessary way to receive the content on the web, which was words. That will be less true in the future as more of the content of the worldwide web is images, not words. Making use of a traditional computer required literacy; the same is not true of a smartphone. Apparently, those images can be powerful in shaping public opinion, and perhaps policy outcomes, both domestically and internationally even if, still, most of the posited effects are anecdotal. Consider the effects of the pictures from Abu Graibh, or the YouTube video of a US helicopter strike apparently killing civilians in Baghdad in 2007.[28] Compare those with the leak through WikiLeaks of 92,000 US military documents pertaining to Afghanistan.[29] If the latter seemed to have less impact, that probably was mostly because the main storylines—collateral damage and Pakistani complicity with the Taliban—were familiar (the former in part because of previous images). But the documents were just that, words, and thus a quintessential newspaper story, not images that might go viral on the web. Furthermore, both

[28] The video is available at http://www.youtube.com/verify_age?next_url=http percent3A// www.youtube.com/watch percent3Fv percent3Dis9sxRfU–ik.

[29] For excerpts and analysis, see the *New York Times,* July 26, 2010.

the WikiLeaks warlogs and the US State Department cables contained simply far too many words, an indigestible information mass demanding extensive decoding and contextualization to become useful. Transparency here created the same effect of massive information overload as intelligence had experienced from the 1960s onward due to the focus on collection (see Chapter 3).

Perhaps because of the newness of social networking, it is not clear how important these media will be in organizing, for instance, anti-government groups. In Iran and other countries, they have been the latest information technology (IT) innovation in organizing in a line that runs from Ayatollah Khomeini's audiocassette speeches smuggled into Iran in the 1970s through the fax machines that played a role in Eastern Europe's transition later. In crisis periods, like the one after the Iranian elections, social media had the attractions of speed and relatively anonymity that offered protection from government retaliation. But they also carried the challenge of validating who was real and who was not (and who was a government agent).

For policy, social media thus far are yet another means for governments to communicate with their constituents and for those constituents to organize to oppose or influence their governments. The impact of social media on commerce is visible and striking. On the political and social side, as opposed to the commercial, it remains unclear whether the uses of social media are in fact effective at increasing participation, increasing knowledge, holding government to account, or overcoming collective action problems. In large part this uncertainty results from the empirically tenuous tie between public opinion and policy, and concerns about selection bias—that is, various types of political participation are generally understood to arise from a broad range of individual-based influences, for example, gender, age, income, education, and so forth. The extrapolation is that online activity is undertaken predominantly by those already politically engaged offline.

So far, the use of new media by governments, what has come to be called e-governance, is oriented largely around soliciting information from citizens. New technology platforms allow this communication to occur in real time and with low transaction costs—complaints are automatically sent to the proper department without use of a phone operator, and stakeholder opinions about policy can be received without the inefficiencies of large meetings. Such initiatives as the Peer-to-Patent program, in which an "open review network of external, self-selected, volunteer scientists and technologists provide expertise to patent examiners on whether a patent application represents a truly new and non-obvious invention," are suggestive of the promise of e-governance uses of social media.[30]

---

[30] Michael Lennon and Gary Berg-Cross, "Toward a High Performing Open Government," *The Public Manager* (Winter 2010).

However, for policy the process may be slower than expected. Entrenched bureaucracies struggle to adapt to this new type and pace of interaction with the public. Managers will need to review and in many cases to change practices and policies that served tolerably well within a hierarchical, agency-as-provider-but-not-recipient model, and proper incentives to encourage such collaboration will need to be put in place.[31] Optimism about the benefits of citizen collaboration in the regulatory and policymaking processes enabled by open government should also be tempered by acknowledging that more participation may introduce costs. Many observers, for example, fear that the new world of electronic mass submissions will overwhelm and delay agencies with limited resources. While concrete examples do not exist, the concerns extend to the prospect that e-government might introduce more, and more intractable, conflicts of interest into the process that can slow rather than speed deliberation.[32] Moreover, the fact that e-government and e-government infrastructures are web-based may mean that the conditions reflected are those only of the populations that have ready access to broadband. Indeed, a 2010 Pew study of the United States found that it is the white, affluent, and well-educated that are most likely to access government information, to use government services, and to participate in collaborative enterprises online.[33]

All these considerations reinforce the proposition that for policy, the implications of transparency and new technology will be more evolutionary than revolutionary. Those technologies will make it easier for citizens to question their government, and perhaps to oppose or even seek to overthrow it. For policy analysis, those technologies will facilitate what transparency is requiring— what Klitgaard calls *convening*. Convening is a long way from "*the* study for *the* decision." In Klitgaard's words,

> Convenings refers to the bringing together of stakeholders to work together on complex issues. Those convened have different if overlapping objectives, different if sometimes overlapping capabilities, and different if overlapping information about the state of the world and about if-then relationships (such as treatment effects). The stakeholders are strategically connected, in the sense that what one party does

---

[31] "Defining Gov 2.0 and Open Government," Alex Howard, January 5, 2011. Available, as of July 22, 2014, at http://gov20.govfresh.com/social-media-fastfwd-defining-gov-2-0-and-o pen-government-in-2011/.

[32] See S. W. Shulman, *The Internet Still Might (but Probably Won't) Change Everything: Stakeholder Views on the Future of Electronic Rulemaking* (Pittsburgh: University of Pittsburgh, University Center for Social and Urban Research, 2004).

[33] Aaron Whitman Smith, *Government Online: The Internet Gives Citizens New Paths to Government Services and Information* (Washington, D.C.: Pew Internet & American Life Project, 2010).

often affects the outcomes of what other parties do. They are not fully aware of each other's objectives, capabilities, or information sets; they do not fully understand their strategic interrelations. It may also be the case that no one can understand all those things, in the sense that the stakeholders (along with the environment around them) form a complex system.[34]

Conceived in this way, convenings are also suggestive for intelligence. They use policy analysis as a medium, much as intelligence analysis might be used, and the process may help stakeholders better understand their own and one another's objectives. The process may enrich participants' sense of available alternatives. It may improve their sense of conditional probabilities, the "if-then" notions of probability, risk, and unforeseen consequences in policy alternatives. And it may help them think more strategically, including about their future relations with each other.

Characterizing the future world of which social media are the leading wave begins to demonstrate just how dramatic the change will be for intelligence. That future world will be one of ubiquitous sensing and transparency. For the former, think of those closed circuit TVs (CCTVs), which number as many as 420,000 in metropolitan London alone, or one for every 14 people.[35] Transparency will be driven by those ubiquitous sensors and their location awareness, by the increasing computational power of devices, and by increases in communication and connectivity. Converged sources have been the buzz word for nearly a generation but only became a reality in recent years with the smartest of smartphones, not to mention the iPad and its kin. Those personal mobile devices indicate *who their users are; where and when they are, have been, and will be located; what they are doing, have been doing, and will be doing; with whom or what they are interacting; along with characteristics of their surrounding environment.* At the same time, threat signatures and footprints will continue to shrink—notice the challenge of attributing cyberattacks. Background noise will continue to rise. Finally, these technological capabilities will continue to proliferate as cutting-edge research and technology go global. If it is not always the case that the bad guys can do everything the good guys can, the gap between the two cannot safely be counted on to be very long.

A series of RAND projects over the last several years has looked at the use of social media by American intelligence agencies. The subjects have been *external* social media for both intelligence analysis and operations and *internal* versions

[34] Klitgaard, "Policy Analysis and Evaluation 2.0," p. 29.
[35] To be sure, these estimates should be treated with caution, for they tend to be extrapolations of counts in a much smaller area.

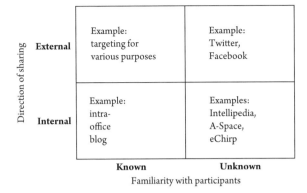

Figure 6.3 Four Kinds of Interaction through Social Media. Source: This figure is drawn from Mark Drapeau and Linton, *Social Software and National Security: An Initial Net Assessment* (Washington, DC: Center for Technology and National Security Policy, National Defense University, 2009), p. 6.

of those tools designed to enhance collaboration within and across agencies. This section focuses on the external uses; Chapter 7 turns in more detail to social media as tools for internal collaboration. Intelligence interactions using social media might be divided into four functions, as shown in Figure 6.3.

The functions are based on whether the participants are known or unknown to the official engaging in the interaction, and whether the purpose is communication internal to the agency or community, or external. Plainly, the distinctions are not discrete but rather continuums, matters of more or less. The US intelligence community's internal tools—Intellipedia, A-Space, and eChirp, for instance—might be put in the lower right quadrant of the figure because most of the participants may be unknown to each other, at least initially. The first is a classified wiki, the second, also highly classified, is intended to help analysts (hence the "A" in its title) from different agencies meet and interact with each other, and the third is a classified version of Twitter. However, if the names of potential collaborators are unknown, salient characteristics are not: they work in or with the intelligence community, and they have security clearances at a known level. These internal uses are "enterprise" uses of social media to enhance collaboration.

The external uses of social media might be divided between targeting and analysis. For targeting, the purpose is less the substance of the posting than the source: as one official put it, "If you are trying to break into the communications of a terrorist, he is likely to have good comsec [communications security]. But he also may have children, who may be on email." So the point is to use social media to target more traditional forms of intelligence collection. Again, the distinction between the two external quadrants may also blur.

A person targeted for recruitment as an informant, for instance, may be known in advance, but may also be unknown initially; rather the target may be identified through analysis of networks on Facebook or other social media.

For analytic purposes, substance is the point. Yet given current technology, because of the huge volume and low reliability, Twitter is useful for intelligence analysis mostly in two circumstances: when information about what is happening is in short supply, and Twitter can supplement it, as was the case during the Mumbai attack of 2008; and when there is no other source of "public" opinion, as in the aftermath of the 2009 Iranian presidential elections. For both uses, validation is key to separating people who might actually be there or are in the opposition from hobbyists halfway around the world, not to mention agents spreading disinformation on behalf of the government. Blogs are somewhat more tractable but still labor intensive. One US intelligence organization produces a weekly report on what's hot in the Chinese blogosphere. It does so by carefully monitoring and selecting blogs. Since Chinese officialdom is known to watch the blogosphere for signs of trouble, the report has become required reading for US officials with China as their beat.

On the operational side, social media can be ignored only at the peril of endangering the cover, perhaps the lives, of intelligence officials. If 20-somethings are not on Facebook or twitter, they only become "conspicuous by their absence." They might as well advertise, "I work for intelligence." Tomorrow if not already today, if officials are under cover, they have to live that cover in virtual worlds as well as physical ones. They have to be prepared to move quickly from one medium to another. Indeed, "official" cover as it has been practiced by the United States and many other nations is already an artifact of the past, though intelligence organizations have not quite realized it. All the elements of transparency—from biometrics through geolocation—mean that any cover will be fleeting, with the possible exceptions of intelligence operatives who really live their cover, being, not just pretending to be, businesspeople or some other professional, and never coming close to any government establishment. And the technologies don't just apply to the future. They also capture the past—for instance, names, passport numbers, and perhaps biometrics like fingerprints that were collected at borders in the past are digitized and thus made searchable in the present and future.

On the analytic side, if the challenges of huge volume and low reliability are already plain, so are some of the possibilities. As tools improve, Twitter and its successors might become sources of if not warning, then at least hints to analysts of where to watch. Social media have the potential to become a repository for collected information, in the process enabling amateur intelligence analysis. Already, for instance, the founder of Greylogic, a noted expert on cyber warfare and a consultant to the US intelligence community, launched a trial balloon on Twitter—called Grey Balloons—seeking volunteers who might spend a few hours per week, unpaid, to help intelligence agencies do their jobs, especially in connecting the dots.[36] Intelligence long has fretted about competition from professionals, like CNN, but now may find itself competing with amateurs for the attention of its customers. It may be asked to vet and validate amateur intelligence efforts, and it may be required to allocate its resources to fill in the gaps not covered by amateur efforts.

Social media are important in and of themselves, but they are also at the cutting-edge of broader changes that are enveloping US intelligence. Social media are active, not passive. They completely blur the distinctions that have been used to organize intelligence—between collector, analyst, and operator, or between producer and consumer. They completely upset existing notions about what intelligence's "products" are. Coupled with smartphones, they will turn any human into a geolocated collector and real-time analyst. They offer enormous promise but also carry large risks and obstacles. Small wonder, then, that a study of social media for the US National Geospatial Intelligence Agency (NGA) found

> enthusiasm mixed with concern. Clear guidance is sparse. Measurement of any benefits these technologies may have brought is not well documented; at the same time there is little quantification of their potential negative impact—instead, a mixture of hand waving and hand wringing.[37]

In one RAND conversation, one of the managers of Intellipedia, that highly classified wiki internal to the US intelligence community, was more positive about the possibilities even as he demonstrated how far from reality those possibilities are. He recognized that it is interaction that drives the possibilities—from asking questions on Twitter to more specific "crowd-sourcing," that is, posing issues for and explicitly seeking help from fellow netizens. A convert, he would take the logic of social media to its conclusion and have the CIA out there, openly, as the CIA, asking questions and seeking help. "Sure," he said, "we'd get lots of disinformation. But we get plenty of that already. And we might find people out there who actually wanted to help us." Seeking that help is one of the keys to reshaping intelligence. So is rethinking what is the "output," *not product* of intelligence. But both are a long way off.

---

[36] See http://twitter.com/greyballoons.; http://intelfusion.net/wordpress/2010/01/12/the‑grey‑balloons‑project‑faq/.

[37] Robert A. Flores and Joe Markowitz, *Social Software—Alternative Publication @ NGA* (Harper's Ferry VA: Pherson Associates, 2009), p. 3.

7

Challenges for Intelligence

The future of intelligence departs from a past of alleged "failure," hence under-performance, giving rise to professional doubts about how the intelligence community thinks it knows what it knows, still less how it adds value to policymaking. Technology and the pace of events are pushing it to think of its work as "outputs," not products, where a conversation may be more important than a paper. In the process, its "customers" are becoming "clients" to be served in ongoing relationships, not serviced at arm's length with products, raising analogies between lawyers and clients, or doctors and patients. Those twin changes raise a host of issues. How is quality assured when interaction with policy is much less formal? Is the "return of endnotes" required, at least for some written outputs? How is trust, both of analysts by their organizations and between analysts and policymakers, established and verified? What is the right professional self-image for analysts; in particular, what does "intelligence advice" mean when intelligence still will be generally enjoined from direct advocacy of policies?

A Postmortem on Postmortems

The following is a list of selected "intelligence failures" by United States intelligence over the last half century:

1940s US intelligence predicts that the Soviet Union is five to ten years away from developing a nuclear weapon. The Soviets detonate a test weapon the following year (1948–1949).

1950s Intelligence reports warn of a Soviet lead over the United States in missiles and bombers. The first US spy satellites put in orbit, beginning in 1960, find no such disparities.

1960s An intelligence estimate says that Soviets are unlikely to position nuclear weapons in Cuba. CIA Director John McCone disagrees and orders more surveillance flights, which soon find signs of missile deployment. Cuban leader Fidel Castro is forced to remove the missiles after President Kennedy orders a US naval blockade of the island (1962).

1970s Persistent shortfalls in estimates of Soviet military capability and expenditure spark "Team B" [insert table note] challenge to the CIA.

1980s US intelligence fails to predict the impending collapse of the Soviet Union.

1990s United Nations (UN) inspectors discover an Iraqi nuclear program that was much more extensive than the CIA had estimated (1991).

India and Pakistan conduct nuclear tests. This testing was not predicted by the CIA (1998).

U.S. warplanes accidentally bomb the Chinese Embassy in Belgrade as a result of erroneous target information provided by the CIA. Three Chinese journalists are killed (1999).

Significant overestimate of the foreign consequences of Y2K (the Millenium Bug) issues (1999).

2000s The CIA fails to forecast 9/11 attacks. It tracks suspected al Qaeda members in Malaysia months before but fails to place Khalid Al-Midhar (one of the 9/11 hijackers) on its terrorist "watch list" (2001).

Iraqi WMD estimate took 20 months to develop and was dead wrong in its assessment of Iraqi WMD.

From the list, it's no wonder that intelligence analysts worry what value they add to the making of policy. Indeed, the question arises, Is intelligence ever right? Yet the better question is, What constitutes "failure" and to what extent can it be avoided? Organizations must learn if they are to compete. If organizations fail to learn, they risk not keeping up with a changing world, all the more so when adversaries are continually trying to frustrate the efforts of US intelligence. But how should they learn, still more to the point, how can they become a learning organization? Those are the questions at issue in this section.

The first half-decade of the 2000s were the season for postmortems of failed intelligence *analyses* (as opposed to intelligence *operations*, which typically were the focus of post-failure assessments in earlier decades). Several of those were detailed and thoughtful.[1] Some of them have been embodied in reform

---

[1] For the United States, the two most detailed are those of the 9/11 Commission and the WMD Commission. Formally, they are, respectively, the Learning from Recent Best Practices Upon the United States, National Commission on Terrorist Attacks upon the United States, *The 9/11 Commission Report: Final Report of the National Commission on Terrorist Attacks Upon the United States*, and L. H. Silberman and C. S. Robb, "The Commission on the Intelligence Capabilities of the United States Regarding Weapons of Mass Destruction: Report to the President of the United States" (Washington, DC, 2005). The British postmortem, the Butler Commission, is Butler Commission, *Review of Intelligence on Weapons of Mass Destruction* (London, 2004).

initiatives in the United States and other countries—for instance, the 2004 US intelligence reform legislation creating the director of national intelligence (DNI).[2] Most of that is all to the good. Nevertheless, as the legal saying has it, hard cases make bad law.[3] The social science equivalent is that single cases—or a linked series of failures in the 9/11 instance—make idiosyncratic lessons; it is all too tempting to conclude that if analysts did x and failed, then they should do *non-x* (or *anti-x*) in order to succeed. All of these postmortems tend to carry the presumption that intelligence analysis is a singular enterprise. Yet it is not. It comprises a variety of purposes and relations to consumers, and thus of data and methods. Intelligence analysis is plural, and so must best practices be plural.

Yet it is far from clear that learning lessons by examining celebrated cases is the best path. The examinations tend to be rare and done in the full glare of publicity—and of political stakes. They often tend to focus on finding guilty villains to blame, as much as improving practice. Even if they do not focus on *who shot—or who missed—John?* they still are methodologically flawed. They focus on a handful of incidents, each with its own peculiarities—but whose lessons are then generalized to inevitably different circumstances. They do tend to assume that if analysts did x and failed, then doing do *non-x* (or *anti-x*) would have produced success, and would do so in future circumstances akin to those examined. By this point, the enquiry is on very thin epistemological ice indeed.

They also focus on *failures*. In addition to demoralizing analysts, that has several consequences. Most obviously, it raises the defenses of those intelligence organizations that feel their copybooks are being graded by the exercise. It is perhaps little surprise, for example, that the first efforts by the US director of national intelligence to create a joint lessons-learned center ran into resistance from virtually all the analytic agencies. Focusing on failures also downplays what might be learned from successes, or even middling outcomes. That preoccupation with major error may also produce a pendulum swing. One among several reasons that US intelligence overestimated Iraq's WMD programs in 2002 was that it had underestimated them in 1990—and been taken to task for doing so.

---

[2] Formally, United States Congress, *Intelligence Reform and Terrorism Prevention Act of 2004 (to Accompany S. 2845)* (Washington, DC, 2004).

[3] The complete quote from which the saying derives is pertinent. "Great cases, like hard cases, make bad law. For great cases are called great, not by reason of their importance in shaping the law of the future, but because of some accident of immediate overwhelming interest which appeals to the feelings and distorts the judgment. These immediate interests exercise a kind of hydraulic pressure which makes what previously was clear seem doubtful, and before which even well settled principles of law will bend."—Oliver Wendell Holmes, dissenting. *Northern Securities Co. v. United States,* 193 U.S. 197, 400–411 (1904).

Perhaps most important, the assessments are seldom very explicit about what constitutes intelligence failure.[4] Not every failure is an intelligence failure; in principle, there could be intelligence successes that were followed or accompanied by policy failures. A dramatic example is the US National Intelligence Estimate on Yugoslavia in the autumn of 1990, which predicted Yugoslavia's tragedy with a prescience that is awe-inspiring.[5] It concluded that Yugoslavia's breakup was inevitable. The breakup would be violent, and the conflict might expand to spill into adjacent regions. Yet the estimate had no impact on policy whatsoever. None. Senior policymakers didn't believe it, or were distracted by the impending collapse of the Soviet Union, or didn't believe they could do anything about it. To the extent that policy officers saw and digested the estimate, intelligence could not be said to have failed. To qualify as an intelligence failure, flawed intelligence analysis has to be seen and acted on by policymakers, leading to a failure. There has to be a decent case that better intelligence would have induced policy officials to take another course, one that was likely to have led to a more successful policy. By that definition, intelligence on Iraqi WMD in 2002 surely was flawed but may not qualify as an intelligence failure to the extent that a better estimate, within the bounds of what was possible, probably would not have changed the policy outcome.

In these circumstances, most of the quick lessons from the recent postmortems are apt. But they tend to be relatively superficial, reached by wise people who usually are amateurs in the esoterica of the trade (because the experts are likely to be seen as biased, even responsible for the failures being autopsied). They are in the nature of reminders that analysts might tape to their computers, less *lessons* than good guidance that is too easily forgotten. After the fall of the Shah in Iran, both intelligence and policy in the United States reflected the big lesson from the case—don't assume the Shah understands his politics any better than US intelligence does—and both applied that conclusion to the next similar case at hand, the fall of Ferdinand Marcos in the Philippines. In that event, the *lesson* produced a success.

By the same token, intelligence took on board the headline from the Indian nuclear test postmortem—all politics may be local but no less important for it, so take seriously what politicians actually say they will do. That lesson may not have been applied forcefully enough to Osama bin Laden, but intelligence analysts were not likely in any case to repeat the mirror-imaging of the Indian

[4] Among many good discussions of this issue, see Stephen Marrin, "Preventing Intelligence Failures by Learning from the Past," *International Journal of Intelligence and CounterIntelligence* 17, no. 4 (2004): 657 ff.

[5] The estimate has been declassified. See National Intelligence Council, *Yugoslavia Transformed*, National Intelligence Estimate (15–90) (1990).

case. Bin Laden presented a different challenge: he was too different even to mirror image; we couldn't imagine how we might act if we were in his shoes.

The headlines from the Iraq WMD case, too, were reminders mostly about good tradecraft, even good social science: validate sources as much as possible, and do contrarian analysis (what's the best case that Saddam has *no* WMD). The report on the biggest of all recent failures, 9/11, is wonderful history with a strong lesson about the importance of sharing and integrating information across US intelligence organizations, not sequestering or cosseting it. Yet even that postmortem cannot escape a certain historical determinism to which case histories are vulnerable: everyone knows how the story ended, and, knowing that, the pointers along the way are painfully obvious. It is easy to underestimate the noise in the data or even, in the 9/11 instance, the good reasons why the CIA and FBI didn't share information very freely or why the FBI didn't go knocking on flight school doors.

Moreover, however valuable these reminders are, they fall short of best practice in learning lessons for intelligence. Warfare, where lessons-learned activities are becoming commonplace, tends to be an episodic activity. By contrast, intelligence, especially in an era of non-state as opposed to state-centric threats, is more continuous. At any one time, it provides estimates of what exists (e.g., how many nuclear weapons does North Korea have?), what will be (e.g., is India planning to test a nuclear weapon?), and what might be (e.g., how would Iran react to the overthrow of Iraq's government?). Intelligence exists to give policymakers reasoned assessments about parameters whose truth-value is not otherwise obvious. Much of Cold War intelligence was about what exists—puzzle solving, looking for additional pieces to fill out a mosaic of understanding whose broad shape was a given. By contrast, intelligence and policy have been engaged at least since 9/11 in a joint and continuing process of trying to understand the terrorist target, in the absence of handy frames of reference.

At any one point in time intelligence agencies are assessing an array of enumerable possibilities, each of which can be assigned a likelihood. A national intelligence estimate is a consolidated likelihood estimate of selected parameters. Such estimates are not (or at least should not be) static. To be valuable, each must be open to adjustment in the face of new information or reconsideration. Sometimes, the new information fixes the estimate firmly (e.g., India tested a nuclear weapon). The 2007 US estimate on Iran's nuclear program, discussed in more detail later, began as good housekeeping, an updating of the community's 2005 conclusion. Before the estimate was completed, though, new information seemed to suggest pretty conclusively that Iran in 2003 had suspended its effort to make weapons while continuing to enrich fissile fuel. More commonly, each event can influence the confidence with which an estimate is held. The analyst's art, and one where the ability to learn lessons is valuable, is in collecting the right facts, developing or choosing the right rule for integrating these facts, and generating the right conclusions from the combination of facts and rules.

This art can be considered a process, and the goal of a lessons-learned capability is to improve the process continually. Toyota, the car company famous for its success implementing quality control in manufacturing, calls the process of continuous improvement, *kaizen*, which is critical to its success.[6]

How does such a process work? Consider, for instance, an estimate of the likelihood that Iraq had a serious nuclear weapons program. New evidence was then found of Iraqi commerce in aluminum tubes. This evidence had to be interpreted, and once interpreted it should have affected the judgment of whether Iraq had a serious weapons program. Conversely, if another week went by in which weapons inspectors again failed to find a serious weapons program in Iraq, that event should have reduced the belief that such a program exists (by how much is another question). Perhaps needless to add, success at finding the information that would make the greatest potential difference in these estimates is a critical measure of success for the Intelligence Community, but only because it feeds these estimates

There are many formal, mathematical approaches to making inferences.[7] However, they are not panaceas and tend to be more useful in more complex problems involving great uncertainty and multiple variables (e.g., estimates, forecasts, and warning) rather than those involving interpretive reporting of events under way. Human judgment is and will remain the core of the analyst's

[6] Popularized in Masaaki Imai, *Kaizen: The Key to Japan's Competitive Success* (New York: McGraw–Hill, 1986). The Japanese concept of kaizen is directly derived from the principles of designed-in (versus fixed-after-production) quality, customer-centricism, and continuous improvement brought to Japan after World War II by Edwards Deming, an American who worked with the Japanese to recover their industrial capacity. His work did not get traction in the United States until kaizen was promoted here in the 1980s as a Japanese management practice known as Total Quality Management. Deming's approach was to examine and improve the system in which operations take place and not simply to reorganize structure or blame the person. A modern incarnation of this thinking is in Six Sigma programs to reduce product flaws as applied at Motorola and General Electric. For example, see http://www.deming.org/regarding Deming's work and some current resources, and http://www.isixsigma.com/me/six_sigma/ regarding Six Sigma methods. It is also quite relevant to service and intellectual processes, like intelligence, in its reliance on three concepts: products alone are not as important as the process that creates them; the interdependent nature of systems is more important than any one isolated problem, cause, or point solution; and non-judgmental regard for people is essential in that people usually execute as the system directs or allows, people are not sufficient explanation for a problem (or a success), and blaming violates an associated principle against waste.

[7] For instance, Dempster-Shafer theory is a generalization of Bayesian logic that allows analysts to derive degrees of belief in one question from probabilities for a related question—for instance, information provided by an observer or informant whose reliability is subjectively assessed. For a description, see http://www.glennshafer.com/assets/downloads/article48.pdf.

art. Yet there surely is analytic value in at least being explicit about the inputs to an assessment and how they were treated. At a minimum, explicitness permits the validity of such assessments to be scrutinized before third parties. Intelligence analysts should compare their estimates over time. How, for instance, do last month's events alter my estimate about the likelihood of any particular outcome in North Korea?

Having noted such deliberations, it is easier to discover why the processes succeeded or went awry. If they went awry, where? Was it the failure to collect evidence and if so, what sort? Was it the misleading template that was used to process the evidence and, if so, in what way? What unexamined assumptions were made in generating the estimate?[8] Was it the failure to integrate the evidence into the estimate properly, and if so, what kind of failure?

The process might start by making explicit the following issues:

- What is believed to be true and with what probability, or degree of certainty?
- What is believed to be not true, that is, things that are unlikely to happen? (For instance, many intelligence failures arise because possibilities such as the fall of the Soviet Union are simply not given enough credence to be evaluated systematically.)
- What, among the many indicators tracked, is deemed to be more or less important to support or disconfirm a judgment? (Many intelligence failures arise not from failure to predict events per se but the failure to realize the significance—the predictive value—of antecedents or triggers.)
- If a particular estimate is true, what indicators would be observable and supportive, and which disconfirming indicators should not be observable (or have evidence that is not credible)?
- What are the conditional (if-then) assumptions by which evidence is interpreted to support or disconfirm estimates? For example, if rulers are known to commit atrocities, then their national standing will decline. Furthermore, what facts or established principles (e.g., evidence from this situation, the history of other situations, or validated psychological theories of attitude and behavior change) should cause analysts to doubt the validity of these assumptions?
- What rules of thumb are used to assess how much a correct estimate matters in the sense that such a judgment would (1) change policies, programs, or operations if there were better intelligence, and such changes (2) would have made a difference to national security or comparably significant outcomes?

[8] See, for instance, James Dewar et al., *Assumption-Based Planning: A Planning Tool for Very Uncertain Times* (Santa Monica CA: RAND Corporation, 1993).

Explicitness serves two purposes. It would show the continuous nature of the intelligence process, making plain that the process sustains an array of estimates, not unlike currency markets, which are ever-changing around the clock. The more operational reason is to make it easier to review the process by which assessments are generated; explicitness, in that sense, is like the flight data recorder in an aircraft, or the log file on a computer process. That explicitness—if adequately captured as part of the development of an estimate—can be useful even if feedback on many estimates comes long after the fact, if at all.

Interacting with Policy Officials under Transparency

As intelligence struggles to prove its utility to policy, it does so increasingly in a fishbowl. To be sure, not every intelligence assessment leaks, even in the United States. And at the tactical level, leaks matter only if adversaries learn of them in time to take action. Most of the time tactical intelligence, especially on the battlefield, is perishable: either the tank column is where intelligence says it is or it isn't. In either case, the intelligence is only useful for a few minutes, and the source of it is of no concern to the tank commander. This book is about neither leaks nor the perils of over-classification. Suffice it to say, though, that if transparency is increasing, it is still the case that one of the consistent themes of statecraft over the last half century has been governments exaggerating the risks of public disclosure. In the United States, Ervin Griswold, then the solicitor general, argued against publication of the Pentagon Papers in the early 1970s on the grounds that release would cause "grave and immediate damage to the security of the United States." Two decades later, however, he recanted that view in his reflections: "I have never seen any trace of a threat to the national security from the publication" of the Pentagon Papers.[9]

The furor over so-called wikileaks in 2010 tells a similar story. The US government could only bemoan the hemorrhage of literally hundreds of thousands of documents classified up to secret. Yet in fact, the documents show US diplomats doing their job and doing it well, better than many observers, including us, would have expected. They were out on the street; their reports were thoughtful, and candid, and the writing sometimes verged on the elegant. The cables display a kind of political reporting that we feared had become a lost art given both modern communications technology and the pressure the State Department felt to cover many new posts with not much more money in the aftermath of

---

[9] "Secrets Not Worth Keeping: The Courts and Classified Information," *Washington Post*, February 15, 1989, A25.

the Soviet empire's disintegration. Robert Gates, then defense secretary and a wizened Washingtonian, offered a nice antidote to some of the sky-is-falling commentary: "The fact is, governments deal with the United States because it's in their interest, not because they like us, not because they trust us, and not because they believe we can keep secrets. Many governments—some governments deal with us because they fear us, some because they respect us, most because they need us. We are still essentially, as has been said before, the indispensable nation. So other nations will continue to deal with us. They will continue to work with us. We will continue to share sensitive information with one another. Is this embarrassing? Yes. Is it awkward? Yes. Consequences for US foreign policy? I think fairly modest."[10]

Still, increasing transparency—either because secret documents do leak or because senior officials want them to—is part of intelligence's challenges for the future. That challenge is graphically illustrated by the 2007 US National Intelligence Estimate on Iran's nuclear weapons program. That episode was deeply ironic: while those who produced the estimate were and remain proud of the tradecraft that went into it, they were also trapped by their own rules. And the furor over the public release of its Key Judgments left policy officials feeling blindsided. As President George W. Bush himself put it, "The NIE had a big impact—and not a good one."[11] The episode is a cautionary one for a future in which citizens will expect more transparency about why decisions were made, including their basis in intelligence, and leaders of government will be more tempted to use intelligence to justify their decisions.

"We judge with high confidence that in fall 2003, Tehran halted its nuclear weapons program."[12] So declared the first clause of the Key Judgments of the November 2007 US National Intelligence Estimate on Iran's Nuclear Intentions and Capabilities. Done by the National Intelligence Council (NIC), those Key Judgments, or "KJs" in intelligence speak, were declassified and released in December, provoking a firestorm of controversy. The clause seemed to undercut not only any argument for military action against Iran but also the international campaign for sanctions against the country that the Bush administration had been driving. President George W. Bush called the opening "eye-popping," all the more so because it came "despite the fact that Iran was testing missiles that could be used as a delivery system and had announced its resumption of uranium enrichment."[13]

[10] At a Pentagon press briefing, as quoted by Elisabeth Bumiller, "Gates on Leaks, Wiki and Otherwise," *New York Times*, November 30, 2010.

[11] George W. Bush, *Decision Points* (New York: Crow Publishers, 2011), p. 419.

[12] All the quotes from the Key Judgments in this case are from National Intelligence Council, *Iran: Nuclear Intentions and Capabilities.*

[13] Bush, *Decision Points*, p. 418.

The second clause of the KJs added the companion judgment: "we also assess with moderate-to-high confidence that Tehran at a minimum is keeping open the option to develop nuclear weapons," and a footnote to that first sentence sought to clarify that "for the purposes of this Estimate, by 'nuclear weapons program' we mean Iran's nuclear weapon design and weaponization work and covert uranium conversion-related and uranium enrichment-related work; we do not mean Iran's declared civil work related to uranium conversion and enrichment."

Yet both second clause and footnote were lost in the subsequent furor. It was precisely that footnoted "civil" nuclear program that was the target of the administration's campaign lest Iran take itself to the brink of a nuclear weapons capacity through purportedly "peaceful" enrichment programs. A UN Security Council resolution in June 2006 had demanded that Iran stop its enrichment activities, and in December another resolution imposed sanctions on Iran. Any halt in Iran's "nuclear weapons program" did little to ease the policy concerns about the possible military implications of its civilian nuclear program, especially its efforts to enrich uranium. The shouting over that eye-popping first clause drowned out the more nuanced findings contained in the balance of the estimate.

The controversy over the estimate was rife with ironies. For those taken aback by that first clause, it was ironic that the estimate attributed the Iranian decision to halt its nuclear weapons program precisely to the international pressure that the conclusion seemed to undercut: "Our assessment that Iran halted the program in 2003 primarily in response to international pressure indicates Tehran's decisions are guided by a cost-benefit approach rather than a rush to a weapon irrespective of the political, economic, and military costs." While the drafters intended that conclusion as positive—diplomacy works— that is not the spin the story acquired.

For the National Intelligence Council and the intelligence community, the immediate irony of the controversy was that the estimate, and the key judgments, were meticulous in many respects, and NIC leaders had worked hard to improve both the process and product of NIEs after the disaster of the October 2002 estimate about Saddam Hussein's weapons of mass destruction. The Key Judgments reproduced a text box from the estimate that carefully explained what the NIC meant by such words of subjective probability as "likely" or "probably" or "almost certainly," as well as judgments like "high confidence." Beyond more clarity in language, the NIC had also sought more rigor in the process, especially by requiring formal reviews by the major collectors of the sources included in the estimate. And, the furor notwithstanding, the primary findings of the 2007 NIE were neither retracted nor superseded, and were in fact reiterated by senior intelligence officials, including the director of national intelligence, many times through early 2012. Some new information, and new scrubs of older information tended to confirm the judgment.

In retrospect, the root of the furor was the presumption of secrecy. The director of national intelligence, Michael McConnell, had promised Congress the estimate by the end of November, and the National Intelligence Board (NIB)—the heads of the various intelligence agencies—met on November 27. The meeting began with an explicit decision not to declassify and release either the estimate or its KJs. An October memorandum from DNI McConnell had set as policy that KJs should not be declassified, and he had made that point in speaking to journalists two weeks before the NIB meeting.[14] Thus, the meeting proceeded on the assumption that what was being reviewed was not a public document but rather a classified one intended for senior policymakers who understood the issues well. On the whole, the NIB regarded the draft estimate as reconfirming previous estimates—with one very significant exception. That exception was the halt in the weaponization program in 2003. The board felt that judgment was so important that it should be the lead sentence, followed immediately by the companion judgment that, at a minimum, Iran was keeping its options open to develop nuclear weapons. Calling attention to a changed assessment was also consistent with the new requirements spelled out in Intelligence Community Directive 203: Analytic Standards.[15]

The president was briefed on the approved estimate November 28, 2007 and it was delivered to the executive branch and to Congress on Saturday, December 1. Critically, notwithstanding McConnell's October policy and the NIB decision, the president decided over the weekend to declassify the Key Judgments. Two lines of argument drove that decision. One could be summarized, in Fingar's words, as "because it is the right thing to do."[16] Because the United States had for years used intelligence assessments in seeking to persuade other nations to act to prevent Iran from getting the bomb, it had some

[14] This account of the meeting is based on Thomas Fingar, *Reducing Uncertainty* (Stanford CA: Stanford University Press, 2011), p. 120. For the McConnell memorandum see Michael McConnell, "Memorandum: Guidance on Declassification of National Intelligence Estimate Key Judgments," Director of National Intelligence, 2007. For McConnell's quote to the press, see Harris, cited above. McConnell's quote to the press, see Shane Harris, "The Other About-Face on Iran, *National Journal*, December 14, 2007, accessed February 21, 2013, http://shaneharris.com/magazinestories/other-about-face-on-iran/

[15] That directive, effective 21 June 2007, is the following from the same author: Michael McConnell, "Intelligence Community Directive Number 203" (Washington DC: Office of the Director of National Intelligence, 2007).

[16] Fingar, cited above, p. 121.

responsibility to tell others that it had changed its assessment about one key part of Iran's nuclear program.

The other argument was less high-minded and more low-down Washington. In the president's words: "As much as I disliked the idea, I decided to declassify the key findings so that we could shape the news stories with the facts."[17] Or as Vice President Cheney put it to Politico on December 5: "There was a general belief—that we all shared—that it was important to put it out, that it was not likely to stay classified for long, anyway. Everything leaks."[18] From the perspective of Stephen Hadley, the president's national security advisor, the "2005 NIE and its conclusions were on the public record. Even if the new estimate didn't immediately leak, members of Congress were bound to compare it with the 2005 version, provoking charges that the administration was 'withholding information.'"[19] The declassified KJs were released on Monday, December 3, 2007.

In the ensuing public debate, the first clause of the KJs dominated everything else, and people tailored it to fit their particular cloth. Iran's president, Mahmoud Ahmadinejad, was jubilant and immediately called the NIE a "great victory" for his country.[20] President Bush noted that momentum for fresh sanctions faded among the Europeans, Russians, and Chinese, and he quoted *New York Times* journalist David Sanger about the paradox of the estimate:

> The new intelligence estimate relieved the international pressure on Iran—the same pressure that the document itself claimed had successfully forced the country to suspend its weapons ambitions.[21]

Administration critics seized on the estimate as evidence that the administration had hyped the Iranian threat, just as it had the Iraqi threat in the runup to war.[22] The president summed up his own puzzlement:

> I don't know why the NIE was written the way it was. I wonder if the intelligence community was trying so hard to avoid repeating

---

[17] Bush, *Decision Points*, pg. 419.

[18] As quoted in Harris, cited above.

[19] Interview by author (Treverton), Stephen Hadley (2012).

[20] As quoted in "As the Enrichment Machines Spin On," *The Economist* (2008), http://www.economist.com/node/10601584.

[21] Bush, *Decision Points*, p. 419. The Sanger quote is also reproduced in David E. Sanger, *The Inheritance: The World Obama Confronts and the Challenges to American Power* (New York: Random House, 2010), p. 24.

[22] See Robert S. Litwak, "Living with Ambiguity: Nuclear Deals with Iran and North Korea," *Survival* 50, no. 1 (2008), 91–118.

its mistake on Iraq that it had underestimated the threat from Iran. I certainly hoped intelligence analysts weren't trying to influence policy. Whatever the explanation, the NIE had a big impact—and not a good one.[23]

For Hadley, the outcome was a kind of Greek tragedy. From his perspective, while the NIC had been indicating that it might change its 2005 view on the weaponization program, that conclusion was finalized only a week before the estimate's release.[24] From that point on, it was indeed

> a Greek tragedy, one that couldn't be avoided. The document was not written to be public. So Mike [McConnell] comes in with the estimate and the change of view from 2005. He says this can't be made public. But the problem was that the 2005 conclusion was on the public record, so when the estimate went to the Hill, there were bound to cries that the administration was withholding evidence, that it was again trying to manipulate public opinion. So the key judgments have to be made public. Mike takes the document away and comes back with very minor changes, the proverbial "happy changed to glad," because the NIE was approved as written. Then it comes to me. I'm caught. I can't rewrite it because then Congress would compare the public version with the classified one, and the manipulation charge would be raised again. But if the KJs had been written from the beginning as a public document, they would have been written very differently.[25]

Surely, the presumption of secrecy was the root of trouble in the case. In retrospect it is hard to understand that presumption. After all, Congress had requested the estimate in the first place. President Bush decided to release the KJs because the NIC's 2005 findings were on the record, and so simple honesty argued for releasing the new view. Moreover, the new finding was bound to be compared to the old, raising charges of manipulating information. Still, the estimate had been prepared under the DNI's decision that none of it—or any future NIEs—would be made public, and good tradecraft dictated that the Key Judgments parallel the estimate as closely as possible. They would have been written very differently for a public audience not steeped in the substance

[23] Bush, *Decision Points*, p. 419.

[24] Stephen Hadley, "Press Briefing by National Security Advisor Stephen Hadley," (December 3, 2007), Washington, D.C.

[25] Interview: Stephen Hadley.

but with axes to grind. The hasty release also preempted a careful diplomatic and intelligence roll-out of the new assessment.

With the benefit of hindsight, the scope note reproduced in the KJs should have opened with a flashing light warning of what the estimate was *not* about—Iran's "civil" nuclear programs that had earned it censure from the International Atomic Energy Agency and sanctions from the United Nations. Instead, while the note detailed the time frame and questions for *its* subject—Iran's nuclear weaponization and related enrichment—it left itself open to a broader interpretation by advertising itself as an "assessment of Iranian nuclear intentions and capabilities." Lots of NIEs have been produced since and remained secret. Yet in the crunch, with several precedents, it will be hard for administrations to resist releasing parts of NIEs, all the harder because NIEs, as assessments agreed to by all the intelligence agencies, acquire a document of record quality. And sometimes, as with the October 2002 NIE on weapons of mass destruction in Iraq, it will be administrations that take the lead in getting "secret" intelligence assessments out in public.

Increasing transparency is changing the relationship between intelligence and policy, perhaps most visibly in the United States. Here intelligence rightly cherishes its independence from administrations in power, and there is a need for some consistent channel—perhaps between the DNI or NIC chair and the national security advisor—to give administrations warning of what subjects are being assessed and what assessments are emerging. In this instance, process was the enemy of warning; whatever advance notice policy officials might have had that the NIC view would change wasn't really actionable until the estimate was formally approved.

New Tools, New Collaboration—Including with "Clients"

The broad implications of new information technologies, especially what are labeled—not too helpfully—"social media" were laid out in the previous chapter. In the United States, as with the intelligence community's exploitation of *external* social media, the process of adopting *internal* collaborative tools also has been bottom-up. Individual officers began using collaborative tools because they found them intriguing and useful. In times of budgetary plenty, there was money to create lots of tools. Indeed, some officers interviewed in the RAND study[26] thought the plenty had encouraged agencies to do their

---

[26] The entirely unclassified version of this RAND study, cited in Chapter 5, is Gregory F. Treverton, *New Tools for Collaboration: The Experience of the U.S. Intelligence Community* (IBM Business of Government, forthcoming).

own things, creating their own tools rather than cooperating in creating collaborative tools. There has not been a strategic view from the perspective of the community enterprise, still less attention by agency seniors, to what kinds of incentives to provide for what kinds of collaboration.

Broadly, the recent history of collaborative tools in the US intelligence community might be divided into two phases: the first, beginning in the 2005 timeframe, might, with some exaggeration for emphasis, be characterized as dominated by *tools,* with the second, more recent phase dominated by *mission.* In the first phase, with all the "cool stuff" coming out of the private sector, intelligence was inspired to build its own counterparts. A-Space is perhaps the clearest example. The implicit assumption was that if the tool was interesting enough, people would use it. A-Space's early managers made that assumption explicit by trying to design not necessarily a destination but at least a way station that officers would want to visit en route to somewhere else. Many of the early adopters hoped for a revolution in the traditional analytic process and even in the way intelligence was disseminated to consumers.

When the revolution did not dawn, some frustration beset the enthusiasts. The goals of the second phase have been more modest. The National Security Agency's NSA's Tapioca (now available on Intelink, an inter-agency system) is perhaps the best example. Its creator sought a virtual counterpart to the physical courtyard that Pixar had constructed at its headquarters—a natural, indeed almost inevitable meeting place as employees went about their daily business. His goal, thus, was "unplanned collaboration," and his animating question from the beginning was "what do I need to do my job here at NSA better?" To that end, where NSA already had tools for particular functions—as it did with Searchlight in looking for expertise—he brought in and embellished the tool. Others spoke of weaving the tools into the fabric of the workplace, not thinking of them as interesting add-ons.

Especially in these circumstances, the label "social media" is not helpful because, as Chapter 6 suggested, the tools are very different. Popular usage tends to lump Twitter and Facebook together when in fact they are very different: the first is open in principle to anyone (who registers) but the second is primarily a means of keeping up with people already known to the user. In that sense, they differ in how "social" they are. Figure 7.1 displays the range of collaborative social media tools in the US intelligence community, both those within agencies and those across them. The within-agency tools are carried on each agency's classified web, and in general they are not accessible to officials from other agencies, even if those officials have the requisite security clearances. The inter-agency tools are carried on what is called "Intelink." Intelink operates at all three levels of classification in the US system—at the SCI, or

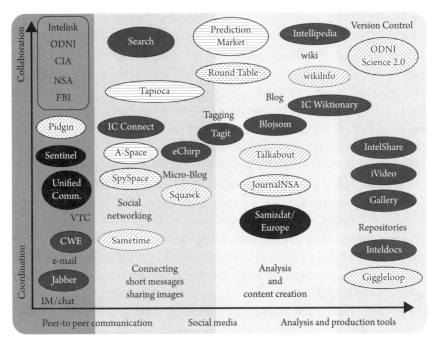

Figure 7.1 Collaborative Media across a Spectrum.

secret compartmented intelligence level; on JWICS; at the secret level on SIPRNet; and at the unclassified level.[27]

On the left-hand side are media that, while often labeled "social," really aren't: instant messaging, Jabber (Yammer) and Sametime (a Lotus Notes instant messenger) in its US community incarnations, are mostly peer-to-peer communications, and in that sense are not very different from email or from the telephone (or letters). In the middle, the tools are more social: eChirp, for instance, the version of Twitter, is accessible to anyone in the community who has SCI clearances and registers. Blogs are relatively accessible. On the right-hand side are tools intended to produce new content. Of those, the oldest is Intellipedia, the community's classified version of Wikipedia. For its sponsors, it suffers by analogy to Wikipedia and gets regarded as only an encyclopedia, when the original hopes for it were to use it to produce intelligence assessments.

The challenge for intelligence in making use of social media, even for internal collaboration, is that those media are the very antithesis of the way intelligence

---

[27] JWICS is Joint Worldwide Intelligence Communications System, interconnected computer networks operating at the Secret Compartmented Intelligence (SCI) level. SIPRNet is Secret Internet Protocol Router Network, a secret-level network widely used by the US military.

has been practiced. Intelligence has been closed and passive; social media are open and active. Intelligence agencies, like most of government, are hierarchical; social media are subversive of hierarchy. Thus, for external purposes, the more open the media, the better for intelligence, for openness creates opportunities both to collect information and target collection operations. However, when intelligence creates versions of those tools for internal purposes, they immediately begin to close—first by being brought entirely behind the security firewall, then often with further restrictions on access. For instance, when a plan was made to turn the US A-Space into i-Space, thus opening it to officials beyond analysts, it foundered, not on technical difficulties but on exemptions from some security rules that the grantors were unwilling to extend to a wider set of officials, regardless of their clearance levels.

In one sense, the challenge that intelligence faces is not so different from that confronted by private industry in trying to increase internal collaboration through social media. Industry too tends for obvious reasons to close down the media when they adapt them. Yet in some instances, companies were able to draw on more open processes in fashioning internal arrangements. For instance, MITRE created "Handshake" primarily as a way for the US Department of Homeland Security to reach out to states and local authorities.[28] It then, however, found the process useful internally as well, and the system continues to be open to outside "friends." It is much harder for intelligence agencies to leverage external connections to build internal ones. Also, rules are more inhibiting for intelligence than for private industry. The plan to convert A-Space foundered on what is called "ORCON"—or originator controlled.[29] That is most notably used by the CIA's clandestine service, which was reluctant to widen the exemptions it had granted for A-Space to a much larger i-Space audience. So too, Intelink has a distribution caveat—NOFORN, meaning that distribution to non US citizens is prohibited. But NSA is very careful about sharing with its "five eyes" partners—Britain, Canada, Australia, and New Zealand—and so NSA officers are unlikely to seek access to Intelink or use it if they have it.

Rules intertwine with organizational culture and process. The RAND studies found that the use of internal collaborative tools, like blogs and Intellipedia, was mostly confined to a small group of enthusiasts. Not only do they *not* feel encouraged by their superiors to use those tools but they also feel that they pay a price for doing so. One CIA analyst, a committed blogger,

---

[28] On MITRE's use of social media, dating back to 1994, see Bill Donaldson et al., "MITRE Corporation: Using Social Technologies to Get Connected," *Ivey Business Journal* (June 13, 2011).

[29] Formally, Dissemination and Extraction of Information Controlled by Originator.

said: "I'm practically un-promoteable." So far, using social media tools to collaborate happens around the edges of existing processes for producing finished intelligence, which remain stovepiped and branded both by agency and author. Indeed, there remains a sense that those who have time to blog can't be first-rate analysts, for if they were, they'd be engaged in the traditional production process. Until that sense changes, the use of collaborative tools among analysts will remain marginal, mostly bubbling up from the bottom.

Collaboration through social media seems likely to remain marginal so long as the existing and traditional processes for producing "finished" intelligence remain in place. Each intelligence agency has a culture and an established workflow for producing finished intelligence. Broadly, the culture and workflows reflect the organization, beliefs, incentives, analytic tradecraft as it is taught and practiced, available tools, and appetite for change of any given agency. One blogger in the RAND study who was skeptical about the traditional process for producing finished intelligence described it as "collect traffic, write paper, publish, repeat." Another emphasized the "push" nature of dissemination by describing it as "fire and forget." Yet as long as the nature of finished intelligence doesn't change, a fair question is how much collaboration makes sense and why.

In imagining what a different future might look like for intelligence analysis and its connection to policy, a prototype produced by the US National Geospatial Intelligence Agency (NGA) called Living Intelligence is provocative. It aims to merge the virtues of crowd-sourcing with agency vetting and to reduce duplication in the process.[30] It would use Google Living Story software, which was developed for a 2009–2011 experiment involving Google plus the *New York Times* and *Washington Post*. Every topic would have its own uniform resource locator (URL). At the top of the page for each "story" would be a summary, and below that a timeline, which the user could move back and forth. On the left side of the page would be the filters, letting users drill down to the level of detail they sought. On the right is a time sequence of important events. In the center is the update stream that keeps track of the entire story. Once a user has read a piece, that piece grays out, so the user need not read it again. The scheme keeps repetition to a medium. For intelligence, it would help to distinguish between useful tailoring for different audiences and the "stock" story merely repeated.

Finally, the content would be fully vetted by the contributing agencies, thus diminishing the worry that content on collaborative tools is second-rate or less reliable. Using WordPress and MediaWiki (the same software used by Wikipedia and Intellipedia), the page would use grayed versus lit agency icons,

---

[30] For a video explaining the idea, see http://www.youtube.com/watch?v=9ft3BBBg99s&feature=plcp.

plus color coding, to make clear which contributions to the topic had been vetted and cleared at which level of the contributing agencies. Both the software programs permit geospatial location, so the topic page would add a spatial dimension as well. The hope behind Living Intelligence was that this form of collaboration would encourage agencies to play to their strengths rather than try to do the entire story themselves. In a more distant future, it is possible to imagine policy officials contributing to the story as well, with their additions clearly marked as such; for critical parts of the intelligence mystery—what might drive foreign leaders, for instance—those policy officials often know more than intelligence analysts.

The vision of Living Intelligence opens a much wider set of issues about what intelligence analysts produce and how they interact with policy officials. For all the change in IT, the intelligence community still tends to think of its outputs as a commodity—words on paper or bytes on a computer screen. The language of commodity permeates: those policy officials on the receiving end are still referred to as consumers or, worse, customers. And while the nature of the interaction between intelligence and policy has changed from the days when, figuratively at least in American practice, intelligence analysts threw publications over the transom to policy officers, that interaction is still thought of as pretty standoff-ish lest intelligence become "politicized," the subject of Chapter 8.[31] In fact, the nature of the interaction has changed. For instance, a 2010 study by the CIA's Sherman Kent School of two dozen intelligence successes found that briefings or conversations were the only "delivery" mode present in all cases.[32]

In fact, while the intelligence community still tends to think it is in the publication business, it really is in the client service business. Paraphrasing the title of an important recent article, intelligence officers synthesize for clients, they don't analyze for customers, and intelligence needs to conceive of its business in that way.[33] The "product" is advice, not a commodity. The process needs to be an ongoing relationship, not a transaction. "Client" is not ideal language, but it does connote services provided by more or less equals. It has the flavor of money about it, and client relationships can be episodic, when needed, not

---

[31] Indeed, many young people no doubt have never seen a transom or have any idea what one is!

[32] The title of the report, "Lessons Learned from Intelligence Successes, 1950–2008 (U)," (Kent School Occasional Paper, 2010), and its major findings are not classified, but the report, alas, is, probably because of some of the details of the cases. Another major finding of the paper provides somewhat inadvertent support for the idea of the intelligence-policy relationship as an ongoing, interactive one: in something like three-quarters of the cases, the first CIA conclusion was diffident or wrong.

[33] Josh Kerbel and Anthony Olcott, "Synthesizing with Clients, Not Analyzing for Customers," Studies in Intelligence 54, no. 4 (2010): 11–27.

continual. For those reasons, "partner" would be preferable to client but it is used into triviality.

There are some precedents for the role. When one of us (Treverton) was overseeing the US National Intelligence Council (NIC) estimates, he came to realize that the NIEs, National Intelligence Estimates, were not the NIC's most important product. Those estimates were homework, keeping the NIC in touch with analysts around the intelligence community; they were also calling cards, demonstrating what the NIC could do. But the real product was National Intelligence Officers (NIOs), not NIEs—that is, not paper but people, experts in a position to attend meetings and offer judgment. In conversations, they didn't have to be so conscious of the line between intelligence and policy; they could offer advice. In many respects, those US intelligence officers who deliver the President's Daily Brief (PDB) to senior policymakers, including the president, are also in the client service business. They are as important as the product they deliver. And so they are regarded by their "clients." James Steinberg, the deputy national security advisor in the Clinton administration and later deputy secretary of state in the Obama administration, spoke for many senior U.S. officials:

> I think the most satisfying part was there was a very clear sense through the relationship with the briefer . . . that was a medium through which we could define our interests and areas of concern, and that requests for information, clarification, follow-up could be pursued. And I thought that was very effective . . . we were extremely well served. . .[34]

The intelligence operation cannot truly transform until it reshapes how it considers its *products*. For all the experimenting through the years with video, intelligence's products remain remarkably unchanged: they are primarily *static, branded, and stovepiped*. They are words on a page or bytes on a computer produced within agency stovepipes that give pride of place to subject matter expertise. Indeed, early in the deliberations of the team of wonderful US intelligence officers, they realized that the language of "products" was itself confining because it tended to channel thinking into the familiar grooves—static and a commodity. Thus, they switched to using "outputs" to open up thinking both about what the intelligence community "produces" and how it interacts with policy officials to share the fruits of its work.[35]

---

[34] Quoted in Gregory F. Treverton, "The "First Callers": The President's Daily Brief (PDB) across Three Administrations," (Center for the Study of Intelligence, forthcoming).

[35] The team's work was published as "Probing the Implications of Changing the Outputs of Intelligence: A Report of the 2011 Analyst–IC Associate Teams Program," in *Studies in Intelligence* 56, no. 1 (Extracts, 2012).

By the same token, in principle, social media, especially wikis but perhaps also Facebook and others, would seem the opening wedge for rethinking outputs. Wikis seem tailor-made for intelligence. Not a static product, they are living documents, changed as new evidence surface and new ideas arise. They let experts come together while permitting interested non-experts to challenge views. And throughout, they provide rich metadata about where evidence came from and who added judgment. Yet a recent RAND analysis for the CIA Center for the Study of Intelligence found no instance of wikis being used to produce mainline products. The closest was the CIA's Open Source Works (OSW), and there the wiki is still more a way of warehousing what the organization knows than of producing anything for an external audience. And second thoughts about implications might argue against too-quick an embrace of wikis as production forms—not that such an embrace is likely in any case. There is a legal requirement for documents of record as part of the decision process, but that requirement could presumably be met through technology if the state of the wiki could be reconstructed for any point in time. However, other questions arise, ones whose answer is less obvious than it seems. Who would get to contribute to the wiki? Just analysts working the account? All analysts? Non-analysts as well? If the last, what criteria, if any, would be applied for access? What about cleared outsiders? Or policymakers?

Yet, short of wikis for producing intelligence, future clients of intelligence will want to receive intelligence on their iPads, if they do not already. They will want to have, though iPads or similar technology, the conversation that senior officials have with their PDB briefers, asking questions, getting answers on the spot, or getting additional analysis soon thereafter. Using these new technologies, however, cuts across how most intelligence services do their business. That is most visible in quality control: one of the reasons that intelligence has clung so tenaciously to words on paper or bytes on a screen is that they can be subjected to a careful process of quality control (leave aside for the moment whether those processes produce better documents or just more vanilla ones). Such control is not possible for analysts answering questions by iPad more or less on the fly. Those PDB briefers prepare to answer the most probable questions, and come to the briefings with notes in order to do so. A less formal process would require empowering analysts. The quality assurance would rest on the people, not their products. And those people would also be the outputs of intelligence.

A pioneering article by Calvin Andrus almost a decade ago described the role of the analyst by analogy to the change on the battlefield.[36] Intelligence

---

[36] Calvin Andrus, "Toward a Complex Adaptive Intelligence Community: The Wiki and the Blog," *Studies in Intelligence* 49, no. 3 (2005), accessed online at https://www.cia.gov/library/center-for-the-study-of-intelligence/csi-publications/csi-studies/studies/vol49no3/html_files/Wik_and_%20Blog_7.htm.

officers, like military units on the battlefield empowered by net-centric warfare, need to be "allowed to react—in independent self-organized ways—to
developments in the National Security environment." To do that,

> they must be more expert in tradecraft. It is this expertise that engen-
> ders the trust required for independent action. Military units know
> the rules of engagement and are thus entrusted to engage in battle . . . .
> Expertise in tradecraft for each intelligence discipline must become
> a constant quest for each officer . . . . Intelligence officers must share
> much more information. Just as military units in the field must know
> where other units are located in geographic space, intelligence ana-
> lysts, for example, must know where their colleagues across the
> Community are located in intellectual space.

Intelligence, and especially intelligence analysis, cannot truly transform
until it reshapes how it considers its *products*.

8

Politicization

Disseminating and Distorting Knowledge

Politicization refers to a process whereby an issue comes to public attention and reaches the political agenda.[1] Defined this way the word is neutral or in many instances has a positive connotation. Politicization in intelligence, on the other hand, has both a different meaning and a distinct negative touch as one of the major causes for failure in the intelligence-policy interface, referring to either various forms of pressure from the policymakers or a bias emerging within intelligence for institutional, ideological, or personal reasons. Politicization in this respect has something to do with bringing an issue to public attention only indirectly, even if the net result can be just that, instead politicization in intelligence refers to a complex interaction between the intelligence and the policy processes, distorting either one of them or both. While politicization in a broader social context can be seen as a part of democracy, it stands out as quite the opposite in intelligence, a distortion of policy, independent analysis, and public perceptions of threats and key security issues.

No single event has served to frame politicization of intelligence in these stark negative terms as much as the conduct of Western intelligence on the assumed Iraqi WMD programs prior to the 2003 war. But politicization is neither a new nor isolated malfunction, nor a concern limited to intelligence organizations and the intelligence-policymaker interface. Politicization, in the negative connotation, is a problem that in various shapes and seriousness transcends virtually all knowledge production, often debated in terms of (insufficient) objectivity and impartiality, and the impact of biases from underlying values and economic or ideological interests. In this broader context, politicization is closely related to over-arching issues of institutional

---

[1] *Oxford Dictionaries* uses as examples the phrases "wage bargaining in the public sector became more politicized" and "we successfully politicized a generation of women."

autonomy and professional ethics, but it also concerns deliberate adjustments, distortions, and frauds. While the latter issues mainly have been highlighted in other domains, in terms of scientific or journalistic misconduct, they also have significance for intelligence and the debate on the causes for and effects of politicization.

In this chapter we discuss the background and definition of politicization in intelligence and science, and suggest an approach for identifying underlying drivers and the dilemmas posed by the cross-pressure between formal expectations, bureaucratic frameworks, and institutional and individual incentives. In the best of worlds there is no such thing as distorting politicization; unfortunately, most knowledge production is carried out somewhere outside the best of worlds.

The Invention of Politicization

In the traditional positivist concept, intelligence was supposed to deliver "facts" and not diverge into assessments and other kind of guesswork, something that policymakers and other recipients many times could do for themselves, and possibly more often *thought* they could manage better on their own. Some of the most skilled analysts in the history of intelligence were indeed themselves sovereigns, policymakers, or military commanders. Intelligence analysis, as we have noted earlier in this book, was both invented and accomplished by those who most badly needed it. This deep-rooted positivist tradition, focusing on "facts," still prevails in the intelligence field, in the organizational structures, the intelligence cycle, and not least in the professional ethos of dealing with facts (in contrast to journalists), the "real world" (in contrast to social constructivists in the academia), and having the duty to deliver these facts about the real world to the customers, even if the facts are unwanted and the "real world" is awful and rejected. Intelligence should, in the positivist tradition, speak truth to power, the only precondition being the ability to secure a three-minute slot and the possession of the Holy Grail of knowledge.

In September 1991, Robert M. Gates, the nominee for the position as director of the Central Intelligence Agency, was questioned by the Select Committee on Intelligence of the US Senate. After decades of turmoil in the relations between the intelligence community and successive administrations over paramount subjects like the Vietnam War, Soviet strategic weapons programs, and the Iran-Contra affair, it hardly surprised anyone that one of the issues raised was that of politicization. Gates began his answer with the reflection that he, as a young analyst, himself had been absolutely convinced that a refusal to accept his analysis "was politically motivated by the people on the seventh floor at the Agency." This is, he continued, an understandable reaction if your work comes back "and it's got criticism written all over." Politically motivated rejection is, argued Gates, after all a lot easier to accept than the fact that the work perhaps didn't meet analytic standards.[2]

While not entirely addressing the question fully, Gates nevertheless put his finger on the importance of distinguishing between cases of politicization and the *perception or suspicion* of politicization, with the latter cases presumably far outnumbering the former. But Gates also noted that the perception of politicization was a two-way street; where the analysts saw political pressure and obedient or over-cautious managers, policymakers tended to see an intelligence community where institutions or individuals had their own agenda to influence or alter policy. Politicization, thus, could refer to (inappropriate) policy pressure on the intelligence process or (inappropriate) intelligence influence on policymaking. One of us (Treverton) identified a number of types of the first form of politicization (see Table 8.1).

This leaves us with two problems of definition. The first one is Gates's observation of a distinction between politicization as such and the perception or suspicion of politicization. The other is the meaning of "inappropriate." Intelligence, after all, is precisely about influencing policy. Without this ambition the whole undertaking degenerates into knowledge production for its own sake.[3] While this is well known for intelligence organizations in decline, there are also numerous cases of powerful intelligence institutions losing sight of policy and degenerating into irrelevance.[4] The other extreme is when intelligence becomes a part of policy, or even replaces policy as sometimes is the case under dictatorship when the heads of a powerful intelligence or security apparatus simply can overthrow a government. Under such conditions everything in the intelligence domain tends to become thoroughly politicized, and the leeway for unbiased intelligence becomes limited or vanishes altogether.

[2] *Nomination of Robert M. Gates, Hearings before the Select Committee on Intelligence of the United States Senate* (Washington, DC: US Government Printing Office, 1992), pp. 510–511.

[3] Mark M. Lowenthal, "A Disputation on Intelligence Reform and Analysis: My 18 Theses," *International Journal of Intelligence and CounterIntelligence* 26, no. 1 (2013): 31–37. Lowenthal states as his first thesis that the intelligence community exists primarily to provide analysis to policymakers and that it has no meaning if it does not act in this supportive role. It has, Lowenthal underlines, no independent self-sustaining function.

[4] The best-described case is perhaps the West German Bundesnachrichtendienst under the long reign of General Gerhard Gehlen; for a brief summary, see Laqueur, *A World of Secrets: The Uses and Limits of Intelligence* (London: Weidenfeld and Nicolson, 1985), pp. 212–219. On the history of Bundesnachrichtendienst (BND), see Hermann Zolling and Heinz Höhne, *Pullach Intern: General Gehlen Und Die Geschichte Des Bundesnachrichtendienstes* (Hamburg: Hoffmann und Campe, 1971) and Erich Schmidt–Eenboom, *Der Schattenkrieger: Klaus Kinkel Und Der BND* (Dusseldorf: ECON, 1995).

Table 8.1 **Defining Politicization**

Type	Description	Ways to Mitigate
Direct pressure from policy	Policy officials intervene directly to affect analytic conclusion.	Rare but can be subtle—logic is to insulate intelligence.
"House" view	Analytic office has developed strong view over time, heresy discouraged.	Changed nature of target helps, along with need for wide variety of methods and alternative analyses. NIE-like process can also help across agencies.
"Cherry picking"	Policy officials see a range of assessments and pick their favorite.	Better vetting of sources, NIE-like process to confront views.
Question asking	How the question is framed, by intelligence or policy, affects the answer.	Logic suggests *closer* relations between intelligence and policy to define question, along with contrarian question-asking by intelligence.
Shared "mindset"	Intelligence and policy share strong presumptions.	Very hard—requires new evidence or alternative arguments.

Source: Treverton, *Intelligence for an Age of Terror* (New York: Cambridge University Press, 2009), p. 175.

Between these two extremes of irrelevance and merging with power lies the complex web of constitutional, professional, and ethical balances that constitutes the core of the politicization issue, not only in intelligence. This complicated terrain is explored in more detail in a chapter by Treverton.[5]

[5] See Gregory F. Treverton, "Intelligence Analysis: Between 'Politicization' and Irrelevance," in *Analyzing Intelligence: Origins, Obstacles, and Innovations,* ed. Roger Z. George and James B. Bruce (Washington: Georgetown University Press, 2008).

It could rightly be argued that the concept of politicized intelligence in its present form is a US invention, one stemming from the specific structure and role of the US intelligence community in the recurring struggles over strategic issues in defense and foreign policy during the Cold War.[6] The focus on strategic armament and deterrence meant that intelligence played a crucial role in the formulation and implementation of policy. This resulted in prolonged and recurring turf battles within the compartmentalized US intelligence system over the assessment of Soviet capabilities and intentions.[7] This is not to say that the development of the US strategic nuclear forces and individual programs were intelligence-based or intelligence-driven, as a large literature on various aspects of the military-industrial complex indicates a predominant role for powerful structural factors. But on the policy level, intelligence was an essential component for the proponents (and sometimes opponents) of organizations or individual weapons programs, as illustrated by the diverging interpretations of intelligence on Soviet strategic buildup in the 1950s, the famous "Bomber Gap" and "Missile Gap" controversies.[8] For most of the smaller countries in Western Europe, intelligence did not play this central role during the Cold War, since defense spending and weapons programs were primarily dependent on international commitments and domestic factors. Not even medium powers, let alone the small, had any capacity to counter specific Soviet armament programs.[9]

Yet the turf battles were not only waged by technocrats over the size of budget slices. Beyond the estimates of production rates and missile accuracy lay more complex issues of assessing Soviet long-term goals and underlying intentions, issues that required interpreting the nature of the Soviet state and the layers of

[6] For a systematic comparison between the US and British intelligence systems, see Chapter 1 of Phillip H. J. Davies, *Intelligence and Government in Britain and the United States,* Vol. 1: *The Evolution of the U.S. Intelligence Community* (Santa Barbara, CA: Praeger, 2012).

[7] See John Prados, *The Soviet Estimate: U.S. Intelligence Analysis & Russian Military Strength* (New York: Dial Press, 1982).

[8] For the Bomber and Missile Gap, see Ibid. and *Intentions and Capabilities. Estimates on Soviet Strategic Forces, 1950–1983,* ed. Donald P. Steury (Washington, DC: Center for the Study of Intelligence, Central Intelligence Agency, 1996.).

[9] One example is Swedish defense planning in the late 1960s that simply scaled down the Soviet threat to manageable dimensions, based not on intelligence assessments but on a crude common sense argument: Assuming that the Soviet Union and its allies could not resort to nuclear weapons in the initial stages of a war, their conventional forces had to be concentrated toward the main adversary. Neutral Sweden then only had to bother about the forces left on the margin, hence the term "the Marginal Effect Doctrine"—not an unusual position for small countries trapped between major powers. This was a kind of political pseudo-intelligence assessment, based on the assumptions that the Soviets did not plan to launch nuclear strikes (which was wrong) and that they would use their forces according to Swedish logic (which was even more wrong). Any diverging views that existed within the intelligence communities were effectively deterred from coming out in the open.

its ideology. These estimates were not only politically relevant but also inescapably political and ideological by the very nature of the subject matter. Many of these conflicts within the US intelligence community were focused on the national intelligent estimate process and the changing organizational setting and procedures. Based on the declassified relevant NIEs, Lawrence Freedman has studied the relationship between the estimating process, strategic policy, and the politicization of the intelligence community in the 1960s and 1970s.[10]

Freedman notes that with increasing political polarization during the 1970s over the interpretation of the Soviet threat, intelligence as an independent voice played a diminishing role. The reason, according to Freedman, was the efforts by senior policymakers like Secretary of State Henry Kissinger to get more useful intelligence products, with more focus on raw data and more transparency about areas of disagreement. The result was a paradox where intelligence, instead of becoming more relevant, counted for less in a process where the policymaking clients "were able to assess the material according to their own prejudices and predilections."[11] This was an example of the "cherry-picking" practice. The culmination of this decline in authority came in the famous Team A/Team B exercise, where the CIA was pitted against a coalition of neo-conservative analysts on not very equal terms; Team A followed normal NIE-standard, was heavy footnoted, and contained contrary opinions, something that the Team B report shunned, and where no dissent was allowed.[12]

The German Estimate: Cry Wolf or See No Wolves?

Politicization, however, is neither a new phenomenon nor something associated only with US intelligence, even if the specific circumstances during the Cold War and beyond turned out to be a rather exceptional breeding ground. If we, however, regard politicization as a number of interlocking dilemmas in the production, evaluation, and utilization of intelligence data and estimates, we should expect to find it in all intelligence-driven or intelligence-dependent processes, though influenced by specific historical circumstances along with the prevailing constitutional and cultural context. True, in instances where intelligence is of no or only marginal significance, we will hardly find any traces, as with the case of small-state defense policy mentioned earlier. If there is some element of

---

[10] Lawrence Freedman, "The CIA and the Soviet Threat: The Politicization of Estimates, 1966–1977," *Intelligence and National Security* 12, no. 1 (1997): 122–142.

[11] Ibid., p. 135.

[12] Ibid., pg. 136.

politicization here, it should be sought in such domains as assessment of labor market impact or national and local spinoffs from defense spending.

The historical development of the concept of intelligence also plays a vital role in the rise of politicization. As long as intelligence was conducted in a pre-modern fashion, focused on the collection of certain (secret) pieces of information, there was hardly an arena where politicization could take place. Rather, politicization presupposed the emergence of a strategic planning and decision-making process and the parallel institutions for intelligence assessments. The history of the development of intelligence analysis and that of intelligence politicization are, if not identical, at least closely related.

Wesley Wark studied the British intelligence estimates of German military growth after the Nazi rise to power in 1933 and offers insights into the maze of long-term threat assessments under varying uncertainty and also the elements of politicization in a more compartmentalized, yet smoothly running system. Wark's main finding is that the British intelligence estimates can be separated into four distinct successive phases—periods of secrecy (1933–1935), honeymoon (1935–1936), blindness (1936–1938), and finally war scare and war (1938–1939).[13] As a result, British estimates went from basically ignoring the rise of the German threat, to being relatively relaxed about it based on the assumption of political reconciliation, to a sudden shift toward perceptions of overwhelming German superiority, and then finally to a more balanced view, too late to avoid war, only a second Munich.

The main story is about insufficient intelligence assets and the inability to grasp the magnitude of a shift from the peaceful détente-oriented Weimar Republic to the Nazi Third Reich, but there were elements of outright politicization in one of its standard forms: the ignoring and suppressing of "unwanted" intelligence data and assessments. For the proponents of the policy of laissez-faire, and later of appeasement, indications of a growing threat would serve no purpose other than to undermine policy. This dilemma is perhaps most vividly illustrated in a dramatized version of the debates in the House of Commons between Winston Churchill and his opponent in the government in the movie *The Gathering Storm*. Certainly, the historical Churchill made ample use of intelligence in his campaign against appeasement, and to the extent he could draw from back channels to air intelligence, this did constitute the other side of the outright politicization; in this case it was justified perhaps in a typical Churchillian manner by the fact that he was right and the others were wrong.

However, most of the politicization that appears in Wark's analysis is of a more indirect and subtle character. One example is the prevailing efforts to look

[13] Wesley Wark, *The Ultimate Enemy: British Intelligence and Nazi Germany, 1933–1939* (London: I. B. Tauris, 1985).

for some ground for optimism when all lights flashed red. In drafting the 1939 strategic assessment, the air staff member of the Joint Planning Sub-Committee (JPC) wrote a note to his colleagues arguing that the picture in the draft was too gloomy. And he stressed that there was considerable evidence that "Germany's belt is already as tight as she can bear" and that Germany, through achieving the current advantage in initial military strength, had used up all its hidden resources.[14] As we know today, the remark of the group captain was both right and wrong—right in the sense that Germany indeed did not have the sufficient resources for a prolonged war, and especially not before conquering most of the European continent, but fatally wrong because Germany had no intention of waging such a prolonged war, instead intending to bypass the Maginot Line and the resource threshold with the *Blitzkrieg* concept.

No equivalent to Churchill's broadsides in the House of Commons appeared in the close relations between intelligence and foreign policy. In fact, as Wark notes, at no stage during the 1930s were there any fundamental contradictions between intelligence reporting and the foreign policy of the government.[15] While this could be regarded as final proof for the *lack* of politicization of intelligence, the obvious failure of British foreign policy in face of Nazi expansionism nevertheless indicates something different. Either there was simply a massive or several successive intelligence failures, or intelligence was influenced, if not outright dictated, by the virtue of political necessity. If Churchill's reading of the gathering storm was justified by that fact that he ultimately proved to be right, the docile intelligence assessments in themselves raise questions since appeasement ultimately failed. Wark concludes that the main negative impact of intelligence was the tendency to supply assessments that seemed to confirm the present line of policy and thus reassure rather than challenge the fundaments of the dogmatic appeasement policy.[16] Intelligence did not cry wolf in time, simply because policy did not require any such cries; on the contrary, appeasement demanded that there be no wolves in sight.

Perceptions of Politicization and Analytic Self-Deterrence

In 1992, the newly appointed Director of Central Intelligence (DCI) Robert Gates, having passed the Senate hearing, addressed his analysts at the CIA auditorium; he warned against the danger of politicization and urged them to

[14] Ibid., p. 226.
[15] Ibid., p. 235.
[16] Ibid., p. 236.

be true to the ideals of the analyst's profession: "Seeking truth is what we are all about as an institution, as professionals, and as individuals, the possibility— even the perception—that that quest may be tainted deeply troubles us, as it long has and as it should." Gates wanted every agency employee from the DCI down to "demonstrate adherence to the principle of integrity on which objective analysis rests, and civility, which fosters a trusting, creative environment."[17] The speech was made not only against a background of the conflicts related earlier but also in the context of what might be called a widespread self-deterrence within intelligence.

This self-deterrence is also visible as an underlying pattern in the tendency of British intelligence to play down the German estimate in the 1930s, but not in the sense that senior managers within the intelligence system held doubts for themselves or constrained the assessment process. Rather the self-deterrence was internalized to the effect that conclusions that would challenge policy simply were avoided. An effort to identify any manipulation in terms of explicit orders for the intelligence to comply with policy would probably come to nothing, or as Wark summarizes: "If there was any candidate for Churchill's hidden 'hand which intervenes and filters down or withholds intelligence from ministers' it was to be found of the process of analysis within the intelligence community itself, especially in the early years of the 1930s."[18] Nor did investigations two generations later of US and British estimates in the Iraqi WMD affair find evidence of such overt interference by policy in intelligence.

The self-deterrence concerning policy embedded in the analytic process is highlighted by Robert Jervis in his discussion of the failure of US intelligence to foresee the fall of the Shah in the 1979 revolution. In *Why Intelligence Fails*, Jervis reflects over his own in-house CIA investigation shortly after the Iranian revolution, and compares this with the equally important but very different Iraqi-WMD intelligence disaster.[19] In the Iranian case, US intelligence was simply caught ill-prepared, with few analysts assigned, limited linguistic capacity, and over-reliance on intelligence from the regime itself. The agency had few sources within the regime or the relevant clerical opposition. But even if the intelligence process on regime stability drove in low gear, the main obstacle to warning was two filters, one cognitive and the other consisting of self-deterrence for politicization in a period when intelligence authority was at its lowest. The cognitive filter was the unchallenged assumption that the Shah,

---

[17] Robert M. Gates, "Guarding against Politicization," *Studies in Intelligence* 36, no. 1 (1992): 5–13.

[18] Wark, *The Ultimate Enemy*, p. 237.

[19] Jervis, *Why Intelligence Fails: Lessons from the Iranian Revolution and the Iraq War* (Ithaca, NY: Cornell University Press, 2010).

if pushed hard enough by domestic opposition, simply would crack down on it, in much the same way that the Soviet power structure in the last years of the Soviet Union was perceived.[20] In both cases, implosion was simply not imagined and hence no warning indicators were identified. The underlying hypothesis could only be refuted by the regime collapse itself, at a point when such a refutation would be of little value for policy.

In the Iranian case, the cognitive filters, much like the British estimates on the German threat, were connected to an element of self-deterrence, resulting in white spots in the analysis. US policy was intended to influence events (which it as it turned out failed to do), but the intelligence analysts avoided taking this into account in their assessments of the unfolding events, knowing that this would be perceived as intrusive politicization from below. Jervis notes that the Iran analysts accepted this limitation and internalized it: "Indeed when I asked them about this discrepancy, they were startled and said that they had not even noticed it."[21] Politicization, or rather efforts to steer clear of it, can therefore be seen as an important cause for the "blindness for blue" discussed in Chapter 6.

Yet sometimes self-deterrence fails, or backfires in unexpected ways. During the Cold War, Swedish security policy built on the doctrine of Northern Europe as a low-tension area and the Nordic countries as detached from superpower politics and strategies. Norway and Denmark were NATO members but nevertheless observed certain limitations on stationing foreign troops and housing nuclear weapons on their territories. Sweden and Finland were non-aligned, but for Finland the case was complicated by a treaty of "friendship, cooperation and assistance" imposed by the Soviet Union in 1948, stipulating that Finland, after consultations, could receive assistance from the Soviet Union in face of a threat from "Germany or with Germany aligned states." While official Swedish policy always underlined the independent and neutral status of Finland, intelligence assessments and defense planning had to take other scenarios into consideration—if, for instance, Soviet forces were to move, under the political cover of the Treaty, westward toward Sweden. The "blue" dimension meant that the overt rhetoric level had to be kept separated from the secret planning and assessments. Politicization here took the form of an over-arching code of conduct of not rocking the boat, of silently acknowledging certain facts and planning accordingly, but not speaking about them in public.

In January 1975 a major mishap occurred, although with the best of intentions. Influenced by the era of détente and advent of the Helsinki Treaty

---

[20] For US intelligence assessments, incremental changes in the perceptions and the dissenting views, see Center for the Study of Intelligence, *At Cold War's End: US Intelligence on the Soviet Union and Eastern Europe, 1989–1991* (Washington, DC: Central Intelligence Agency, 1999).

[21] Jervis, *Why Intelligence Fails*, p. 20.

highlighting credibility-building measures and transparency, Swedish Defence Intelligence decided to produce an unclassified version of the secret Annual Report. The result was a modest and not very eye-catching stencil paper drawing facts and figures from open sources, augmented by some common sense reasoning about the capacity of NATO and the Warsaw Pact to continue armament programs and in a crisis contingency to move forward reinforcements and logistic support on the Northern Flank.[22] Most of the assessment passed unnoticed, the one exception being the section that dealt with the technical capacity of the Finnish road and railroad system to facilitate a forward deployment of the Soviet forces in the Leningrad military district.

The report would presumably have passed unnoticed had it not been for the part on Finland, which received all the media attention defense intelligence had hoped for, unfortunately in the wrong way. The reaction from Finland was one of irritation, and the Swedish Foreign Ministry was not particularly amused when the media speculated about a rift between foreign policy and the armed forces. No one doubted the figures in the report; the issue at stake was why the supreme commander had made them public at this particular time. As often is the case, the actual explanation was too trivial to be credible: the unclassified report was produced with the same method as other joint reports and assessments, that is, through the compilation of drafts submitted by the various intelligence branches including the one dealing with sea, air, and land communications. Analysts simply rewrote their normal assessment to fit the unclassified format, not realizing that journalists and others who were not frequent readers of the intelligence output would interpret the text in a different way, quoting the Finnish section out of context and presuming a prima facie case of foreign policy politicization.

However, the old saying that a misfortune seldom comes alone proved to be true also on this occasion. The next year a new secret annual report was produced, and since no orders to the contrary had been issued, the analyst staff duly compiled an updated version of the unclassified report. But what about the Finnish railways and roads? They were still running in east-west directions and their capacity had not diminished. Remaining for a short time in the best of worlds, defense intelligence clung to the professional ethos of autonomy and prevailing truth. Said and done, the fateful section was duly reproduced. But remembering the turbulence, the chief of the defense staff took the time to read through the text before it went to print, and very late in the day (literarily) to

---

[22] "Militärpolitik och stridskrafter – läge och tendenser 1975, Överbefälhavaren Specialorientering" (Military Policy and Force Postures—Situation and Tendencies 1975, Supreme Commander Special Report 1975–01–14) (Stockholm: Försvarsstaben, 1975).

his amazement found the unfortunate passage repeated and promptly ordered its removal. As all typists had left for the day, the only option remaining was to simply delete the text, leaving a conspicuous empty space, which journalists immediately discovered just fit the Finnish section in the previous year's report.

This time the speculation was not over politicization from below, but the contrary: the visible evidence of political pressure leading to the censoring of an intelligence product, the incontestable evidence being the demonstrative way in which the armed forces had carried out the order. After this final detour into the conspiracy theories of perceived politicization, the whole enterprise with an unclassified version of the Annual Report was abandoned for good.[23]

The Iraqi WMD-Estimates: Politicization or Indigenous Intelligence Failure?

After 2003, the Iraqi WMD estimates and the issue of politicization have been generally regarded as inseparable. The erroneous estimates, finally confirmed by the findings or rather non-findings of the vast intelligence excavation effected by the Iraqi Study Group, were generally perceived as a clear-cut case of top-down politicization, as US and British intelligence communities simply produced and disseminated the assessments and supporting evidence that the policymakers needed to make their case for war, domestically and internationally.[24] As mentioned earlier, however, both the investigation conducted by the US Senate and the British Hutton and Butler inquiries failed to find any proof that the intelligence process and assessments had been distorted due to pressure from the policymakers.[25] But politicization was here approached in a legalistic fashion, and thus the policymakers were acquitted mainly due to the

[23] The narrative of the episode is based on oral history. The author (Agrell) worked as a junior analyst in Swedish Defence Intelligence at the time. The white pages in the 1976 unclassified report, however, constitute a irrefutable historical remnant of the debacle and speak for themselves.

[24] For a scholarly study making this case, see Liesbeth van der Heide, "Cherry-Picked Intelligence. The Weapons of Mass Destruction Dispositive as a Legitimation for National Security in the Post 9/11 Age," *Historical Social Research* 38, no. 1 (2013): 286–307. Also John N. L. Morrison, "British Intelligence Failures in Iraq," *Intelligence and National Security* 26, no. 4 (2011): 509–520.

[25] United States Congress—Select Committee on Intelligence, *Report on the U.S. Intelligence Community's Prewar Intelligence Assessments on Iraq* (Washington: United States Senate, 2004); B. Hutton and Great Britain, Parliament, House of Commons, *Report of the Inquiry into the Circumstances Surrounding the Death of Dr. David Kelly C.M.G.* (London: Stationery Office, 2004); Butler Commission, *Review of Intelligence on Weapons of Mass Destruction* (London: Stationery Office, 2004).

lack of sufficient evidence of direct involvement in the intelligence process. Especially the US Senate put the blame entirely on the intelligence community, framing the Iraqi estimates as a case of devastating intelligence failure due to flawed analysis and inadequate or ignored quality control.[26]

Few other cases of intelligence failures and possibly no other instance of intelligence-policymaker interaction have been so thoroughly investigated, studied, and debated as the Iraqi WMD case.[27] This in itself constitutes a problem of validity, similar to the focus on intelligence failures discussed in Chapter 7; specific estimates under intense scrutiny can stand out as odder and more controversial if the background "noise" of similar but less know cases is ignored.[28] But, in fact, how wrong were the actual assessments prior to the 2003 war? After waves of scorn over the worst intelligence failure since the 1940s and the conclusion by the US Senate that pre-war intelligence simply had displayed poor tradecraft and insufficient management and failed in virtually every conceivable respect, the question could seem an irrelevant one.[29]

It is not, as we discussed in the previous chapter. Indeed, in his "corrected" version of the October 2002 Special National Intelligence Estimate (SNIE) on the Iraqi WMD programs, Richard K. Betts makes remarkably small changes, mainly by inserting more explicitly descriptions of the uncertainty in the assessments. But the overall conclusion, that Iraq probably was hiding stocks of chemical and biological weapons and was pursuing active development and production programs for these as well as nuclear weapons, remained more or less the same.[30] Betts points out a paradox overlooked in many of the postwar

---

[26] United States Congress—Select Committee on Intelligence, *Report on the U.S. Intelligence Community's Prewar Intelligence Assessments on Iraq.*

[27] Jervis, *Why Intelligence Fails,* and Richard K. Betts, "Two Faces of Intelligence Failure: September 11 and Iraq's Missing WMD," *Political Science Quarterly* 122, no. 4 (2007): 585–606. Also Charles A. Duelfer and Stephen Benedict Dyson, "Chronic Misperception and International Conflict: The US-Iraq Experience," *International Security* 36, no. 1 (2011): 73–100, and Olav Riste, "The Intelligence-Policy Maker Relationship and the Politicization of Intelligence," in *National Intelligence Systems,* ed. Gregory F. Treverton and Wilhelm Agrell (Cambridge: Cambridge University Press, 2009).

[28] Examples of attempts to contextualize the Iraqi estimates can be found in the Butler Report (Butler Commission, *Review of Intelligence on Weapons of Mass Destruction*), comparing Iraq with other cases of proliferation concerning Iran, Libya, North Korea, and Pakistan. Also the second US inquiry on the Iraqi assessments, the Robb-Silberman Commission (*The Commission on the Intelligence Capabilities of the United State Regarding Weapons of Mass Destruction,* Washington D.C: Executive Agency Publications, 2005) discussed this wider intelligence setting.

[29] United States Congress—Select Committee on Intelligence, *Report on the U.S. Intelligence Community's Prewar Intelligence Assessments on Iraq,* pp. 14–29.

[30] Richard K. Betts, *Enemies of Intelligence: Knowledge and Power in American National Security* (New York: Columbia University Press, 2007), pp. 121–123. Betts, "Two Faces of Intelligence Failure," pp. 603–606.

comments on an apparent intelligence failure—that the assessments from a methodological standpoint were not "dead wrong." Intelligence was stuck between the irresponsible standpoint of not offering any judgment at all and the need to take a reasonable stand based on previous knowledge and available intelligence. Indeed, Betts concludes that given the available evidence, "no reasonable analyst" could have concluded in 2002 that Iraq did *not* have weapons of mass destruction.[31]

Jervis, in his efforts to map the causes of the failure—since, after all, no one would argue that the Iraqi estimate is a successful benchmark—writes that the analysis was trapped by the fact that the "truth" as it turned out was so implausible. The eventual findings of the Iraqi Study Group on point after point would have sounded not only improbable but distinctly ridiculous if presented before the war—for instance, that the lack of confirmation that large quantities of anthrax had been destroyed after the 1991 war was due to fear of Saddam's anger if he found out that it had been dumped near one of his palaces.[32] History is often littered with coincidences and trivialities. A fully correct estimate could simply not have been substantiated with the available intelligence, context, and sound reasoning. A wild guess could have got it right, recalled McCone—but would have been useless to policymakers faced with handling the Iraqi quagmire.

So, if the Iraqi estimate was not "dead wrong," why did it nevertheless slide off the mark and overstate certainty, elevating the credibility of intelligence from shaky sources,[33] and not least missing the very possibility of alternative interpretations? The answer is in one sense very simple: the long history of the Iraqi WMD issue, the underestimations prior to the 1991 war, and the continued evasion in dealing with the United Nations Special Commission (UNSCOM) all produced a context, albeit biased, that consisted of an overwhelming list of crimes.[34] Jervis points to the pitfall of believing that scientific methods would make the difference. On the contrary, "intelligence strives to follow scientific method, and every day scientists see results that contradict basic scientific laws, which they react to not by rushing to publish but by throwing out the data because they know it cannot be right."[35] As we have elaborated in Chapter 4, intelligence failures and scientific failures are not all that different and in this respect it is therefore not to

[31] Betts, *Enemies of Intelligence,* pp. 115–116.

[32] Jervis, *Why Intelligence Fails,* p. 147.

[33] The most prominent case was the Iraqi HUMINT source with the ill-fated covername "Curveball." See Bob Drogin, *Curveball: Spies, Lies, and the Con Man Who Caused a War* (New York: Random House, 2007).

[34] Jervis, *Why Intelligence Fails,* pp. 150–153.

[35] Ibid., pp. 149–150. Though, as he remarks: "There is no such thing as 'letting the facts speak for themselves' or drawing inferences without using beliefs about the world, and it is inevitable that the perception and interpretation of new information will be influenced by established ideas."

be expected that analysis by trained academics or the employment of scientific methods would save the day for intelligence analysis when faced with an improbable reality. A science of intelligence would have been just as stuck in epistemological orthodoxies, institutional limits, and individual agendas.

The somewhat surprising outcome of the after-action analysis by both Betts and Jervis is that the unprecedented intelligence failure had less to do with equally unprecedented flaws in the analytic process and more with specific problems associated with the Iraqi case, and the historical setting of 2002–2003. Failure was, to some extent at least, unavoidable, a conclusion we tend to deny as unsatisfactory and apologetic. Intelligence, after all, must be able to deliver. Betts, however, is of another opinion. It was not only the perception of the Iraqi WMD program that was flawed but also the belief that sound intelligence analysis *must* be able to deliver the correct result. And he concludes: "Hindsight inevitably makes most people assume that the only acceptable analysis is one that gives the right answer."[36]

Then what about politicization? Was the US postmortem wrong in putting the entire blame on the intelligence community's poor performance but right in concluding that the failure was not caused by political pressure down the chain of command? Few who have studied the case and the postmortems seem to agree on this point, instead viewing the framing of the postmortems as a part of a "blame game."[37] Both the US Senate and especially the Hutton inquiry approached the issue of politicization from a narrow legalistic perspective, searching for conclusive evidence of improper involvement of the policymakers in the intelligence process and output. Jervis on this ground dismisses the postmortems as "almost as flawed as the original estimates and partly for the same reason: the post mortems neglected social science methods, settled for more intuitive but less adequate ways of thinking, and jumped to plausible, but misleading conclusions."[38] They very much fit the critique of postmortems outlined in Chapter 7. In the case of politicization, this meant that the investigators were simply searching for evidence of the wrong kind of influence. It was hardly orders down the chain of command, but far more the whole collective perception of the coming war and what was expected of—and appreciated from—the intelligence services in the countries trying to assemble an international coalition backed by a UN Security Council resolution.

Politicization in this respect had more to do with the transformation of the producer-customer relationship that occurred after the end of the Cold War: it

[36] Betts, *Enemies of Intelligence*, pp. 122–123.

[37] Davies, *Intelligence and Government in Britain and the United States*, Vol. 1: *The Evolution of the U.S. Intelligence Community*.

[38] Jervis, *Why Intelligence Fails*, p. 123.

had become one in which intelligence services in many instances had to make their work and input relevant.[39] From this perspective, the surge for the Iraqi estimate from spring 2002 was simply an all-time high in customer-pull and producer euphoria. This sudden fading moment of overlapping interests is perhaps best caught in the memo from Tony Blair's press secretary Alastair Campbell to the chairman of the Joint Intelligence Committee (JIC), Mr. Scarlett, in which Campbell expressed his gratitude for the services rendered by "your team" in producing the text for the public dossier to be known as the September Report: "I was pleased to hear from you and your SIS colleagues that, contrary to media reporting today, the intelligence community are taking such a helpful approach to this in going through all the material they have." In the preceding line Campbell sums up the changing role of intelligence: "The media/political judgement will inevitably focus on 'what's new?'"[40]

Science: A Different or Parallel Story?

Politicization in science is not a fundamentally different phenomenon, but it is framed in another way, reflecting the social role of science and scientific institutions in contrast to intelligence institutions that are more directly connected to policy and operations and normally explicitly not tasked with the production of knowledge for its own sake. A core element in the academy-intelligence divide, as discussed in Chapter 2, is the perception in intelligence of *not* being academic, and the corresponding perception in the academy of *not* being some sort of intelligence-affiliated activity. This said, the mechanisms of politicization bear some striking similarities between the two domains, increasing as both areas become more exposed to the glare of media.

In the philosophy of science, autonomy commands the high ground in a way very similar to the traditional positivist ethos in intelligence. So while Karl Popper wrote about the open society and its enemies, Robert Gates warned his analysts to be vigilant against politicization and Richard Betts referred to the "enemies of intelligence."[41] Science and intelligence in this respect stand out in many ways not only as methodological twins but also as somewhat odd and unexpected

[39] For a discussion of this process in the United States, see Gregory F. Treverton, *Reshaping National Intelligence for an Age of Information* (New York: Cambridge University Press, 2003), pp. 197–202.

[40] Hutton and House of Commons, *Report of the Inquiry into the Circumstances Surrounding the Death of Dr. David Kelly C.M.G.*, p. 109.

[41] Karl Raimund Popper, *The Open Society and Its Enemies,* Vol.1: *The Spell of Plato* (1945), and Vol. 2: *The High Tide of Prophecy: Hegel, Marx and the Aftermath* (London: Routledge and Kegan Paul, 1947).

comrades-in-arms fighting a defensive battle to ward off attempts by hostile forces to infiltrate, take control over, and distort the production of knowledge.

The most striking case of politicization of scientific output in contemporary history is probably the rise and eventual fall of "lysenkoism" in Soviet biological research discussed in Chapter 4. The emergence of T. D. Lysenko's quasi-science and its triumph over contemporary genetics represents an unprecedented scientific dispute, even claiming the liberty and lives of some of Lysenko's opponents. Stalinism offered an environment for extreme politicization, where lysenkoism could be portrayed as ideologically correct and the research results of the opponents as heresy. Science was here completely subordinated to the power struggle in a totalitarian state, resulting in a total loss of autonomy and any claims for objectivity or authority.[42] The triumph of lysenkoism was made possible by the hegemony of ideology in the Soviet state but was nevertheless a case of politicization from below. Lysenko did not act on party orders; rather, he was a cunning entrepreneur maneuvering within the system and exploiting the need to find scientific solutions to chronic problem of productivity in Soviet agriculture.

The rise and prolonged survival of lysenkoism in the face of overwhelming international scientific refutation is an extreme case but nevertheless illustrated more general patterns in science policy and a drift toward politicization of the content of science. The success of Lysenko and his followers was based on their claim of being able to deliver improved breeding methods faster and with a greater impact than the proponents of competing theories. In a case where the dispute would concern an issue without direct policy implications, like the continental drift theory, politicization would be far less likely, and the dispute would be contained within the scientific domain.

From this, it follows that the development of postwar science policy, focused on science for societal goals and research financing from governmental agencies, private business, and international organizations, opened up the possibility for various forms of politicization similar to those in the intelligence domain. Lysenko was not a unique case of a scientific entrepreneur climbing up through the system at the expense of opponents and exploiting the expectations of the policymakers and funding agencies. What was so exceptional about Lysenko was that he based his claims on plain research fraud, with manipulated or invented data from his experimental

---

[42] In 1948, the Soviet geneticist Rapoport tried to defend the chromosome theory but was refuted with quotations from a speech by Molotov. Rapoport asked why his opponent thought that Molotov knew more about genetics than he did, a question that led to his expulsion from the Communist Party and dismissal from his post. Z. A. Medvedev and T. D. Lysenko, *The Rise and Fall of T.D. Lysenko* (New York: Columbia University Press, 1969), p. 122.

farm.[43] The far more common case of politicization from below occurs when scientists frame their research in policy-relevant terms, thereby attracting media attention, funding opportunities, and possibilities to influence policy, precisely the kind of distortion that Gates warned against in his speech to the CIA analysts.

The field of gravitation surrounding especially large-scale research funding constitutes a powerful ground for bias, if not in the very results then in the selection of research themes and project design. In intelligence, this bias is to some extent internalized in the concept (or axiom) of the intelligence process, where the intelligence requirements are not decided by the agencies themselves, and certainly not by the individual managers or analysts, but are the prerogative of the superiors, the policymakers, or the operators. Intelligence, as Mark Lowenthal has noted, has no self-sustaining function; it exists only to serve decision makers.[44] It could be argued, however, that the transforming role of intelligence and the increasingly complex problems and targets it must address, as discussed throughout this book, actually constitute a basis for the development of an intelligence equivalent of basic science in some specific fields.[45]

The Potential and Limits of Advice

In his book *Beyond the Ivory Tower*, the British government's chief scientific advisor in the 1960s, Sir Solly Zuckerman, observes that scientists "are poorly equipped, if indeed equipped at all, to predict the social and political repercussions of the application of science and technology."[46] In many instances, scientific advisers nevertheless have ventured out into this domain beyond their competence, often in the belief that they, as scientists, were by definition delivering scientific advice, while in fact they were giving more or less amateurish political advice. A nuclear scientist does not necessary know more about public perceptions of risk or a reasonable trade-off between conflicting policy goals. The former head of Israeli military intelligence, Major General Shlomo Gazit, makes a similar observation regarding intelligence advice to policymakers.

---

[43] Ibid., pp. 233–236.

[44] Mark M. Lowenthal, "A Disputation on Intelligence Reform and Analysis: My 18 Theses," *International Journal of Intelligence and CounterIntelligence* 26, no. 1 (2013): 31–37.

[45] A field where the need for basic research has long been acknowledged is cryptology, where the lack of long-term efforts could result in losses of access that are hard or practically impossible to regain.

[46] Solly Zuckerman, "Beyond the Ivory Tower: The Frontiers of Public and Private Science" (New York: Taplinger, 1971), p. 106.

Intelligence experts tend, according to him, to disregard the complexity of the policy arena and hence the limitations of their own expertise.[47] Politics is neither science nor dispassionate intelligence analysis, and decisions sometimes are taken, and must be taken, contrary to such advice, however well founded.[48]

There seems to be a distinction between the intelligence and scientific domains when it comes to attitudes toward advice. Intelligence is reluctant to be drawn into the political domain, and certainly, at least in Western democracies, policymakers hold a corresponding reluctance or even suspicion. This can, as in the case of the US estimates on Iran prior to the downfall of the Shah, result in a knowledge gap, where crucial aspects disappear. Scientific advice is based on slightly different premises. While policymakers (and their staffs) normally find themselves fully capable of digesting and evaluating the significance of an intelligence output, this is hardly the case with scientific issues. Nuclear energy became one of the first fields where policymakers sought scientific advice, not only in terms of scientific input, but also, and perhaps primarily, to explain the nature of nuclear physics and its implications.

The different premises for scientific advice are also visible in the evolution of climate change as a global policy issue, a process in which scientific expertise not only played a key role in discovering and analyzing the problem but actively "politicized" the climate issue.[49] It was the scientists' growing concern and consensus on the need for imminent political action to prevent a slowly growing threat that elevated the issue. Advice in this context was to place the climate issue on the international policy agenda and keep national politicians committed. In a sense scientists here blurred the dividing line between science, as detached and impartial, and politics, but they did so with reference to a greater common good. The blurring of the dividing line, however, was not unproblematic. The scientist not only found themselves in the role of advisers but also in the role of advocates and in some sense prophets. Gates's words of caution probably would have sounded irrelevant when global survival seemed to be at stake, but the net effects, the rearrangement of arguments, the discarding of alternative hypotheses, and the focus on consensus as a basis for validity—all discussed in Chapter 4—seem altogether too familiar.

[47] Shlomo Gazit, "Intelligence Estimates and the Decision-Maker," in *Leaders and Intelligence*, ed. Michael I. Handel (London: Frank Cass, 1989).

[48] Some elements in British decision making prior to the 2003 Iraqi war suggest that overriding policy concerns, rather than intelligence misperceptions, were the prime movers. A subsequently published memo by the head of the British Secret Intelligence Service from July 2002 indicates an early commitment to close rank with the Americans that, in the opinion of "C," were set on military action. Morrison, "British Intelligence Failures in Iraq," p. 513.

[49] See Spencer R. Weart, *The Discovery of Global Warming* (Cambridge, MA: Harvard University Press, 2003).

Media Attention as the Continuation
of Politicization with Other Means

One important difference between the controversies over politicization in the Cold War and in the runup to the 2003 Iraqi war was the transformed role of intelligence as not only a producer of estimates affecting policy but also, and foremost, as a producer of the intelligence foundation for the public case for war—an issue touched on in the previous chapter. Mr. Campbell's remark to Mr. Scarlett had nothing to do with threat assessments and strategic policymaking as such, for those assessments and decisions had already long been made and on other grounds; intelligence, faulty or not, played an insignificant role in the later stages of the Iraq crisis in 2002–2003.[50] In this respect the worst intelligence failure since Pearl Harbor or the Ardennes offensive in 1944 mattered remarkably little, or as Jervis observes, we cannot take for granted that intelligence explains policy.[51] It was rather the media attention to intelligence that mattered, and that caused the turmoil once the war was over and the weapons of mass destruction were missing, or as the grand old man of British intelligence, Michel Herman, put it: "There have been other intelligence failures but none that in a way has left the British people feeling that it was sold a false parcel of goods."[52]

A number of factors make the case of the Iraqi estimates special or even unique in this respect. The fact that Iraq was completely overrun and occupied denied the benefit of doubt that inaccurate intelligence estimates normally enjoy. Most if not all Western estimates of Soviet capabilities and intentions during the Cold War were probably wrong, at least to some extent. However, it was only rarely that actual events verified the assessments, and even rarer that this verification was done in public, as was the case with the 1962 US NIE on Soviet deployments to Cuba. A second, related factor was the fact that heightened expectations for major revelations in the excavations and an already heated political debate on the *casus belli* made anything but a thorough investigation and open presentation of the findings impossible. A third factor, following from the second, was that the domestic and international public from autumn 2002 had been drawn into the process and singled out as the final customer in an intelligence-declassification-public relations cycle that could not simply be discontinued and denied once the war was on.

The painstaking postmortem of especially the key British unclassified intelligence product, the so-called September Dossier, illustrates the consequences

---

[50] See Chapter 10 of Davies, *Intelligence and Government in Britain and the United States*, Vol. 1.

[51] Jervis, *Why Intelligence Fails*, p. 125.

[52] Interview of Michael Herman, BBC *Panorama*, July 11, 2004 quoted in John L. Morrison, "British Intelligence Failures in Iraq," *Intelligence and National Security* 26, no. 4 (2011): 509–520.

when intelligence is rerouted from the intelligence/policy interface to the policy/media interface. This was not a case of JIC informing the government, its main task for over 60 years, as Tony Blair underlined in the first paragraph of his foreword to the dossier.[53] For the first time this trusted intelligence provider and adviser was reaching out with an unclassified version of the classified intelligence assessments. But, as the Butler Inquiry observed, the JIC put its reputation at stake when taking responsibility for the intelligence content of something that in effect was a skillfully drafted policy product with the result that "more weight was placed on intelligence than it could bear."[54]

The ill-fated September Dossier bears some resemblance to the efforts of Swedish defense intelligence to produce an unclassified assessment in the 1970s, though the latter was by no means a government-approved, let alone requested, undertaking. The main problem in both cases was not errors or omissions when translating from secret to open, but the unforeseen interpretation gap when an intelligence product was taken out of the dissemination flow and released to readers lacking this context as well as the ability to read intelligence prose. In this, it was very much like the US NIE on Iran's nuclear program. The use of intelligence to make the public case for war against Iraq could constitute a form of politicization where intelligence, as was obviously the case in Britain, somewhat reluctantly was compelled to leave the expert role and venture out in an area of public controversy.[55] However, it also heralded a changing role of intelligence in which media not only is a channel for dissemination but also an arena that intelligence has to take into consideration and that is transforming into a parallel oversight actor, especially after the rather poor performance of independent journalism between 9/11 and the Iraq War.[56]

The media exposure of science has, as underlined by most authors in the field, also seen a shift from the 1970s and onward from a predominantly uncritical reporting of exciting discoveries to a critical scrutinizing of scientific results, controversies, and performance failures.[57] Science—and later

---

[53] Joint Intelligence Committee, *Iraq's Weapons of Mass Destruction: The Assessment of the British Government* (London: Stationery Office, 2002).

[54] Butler Commission, *Review of Intelligence on Weapons of Mass Destruction*, p. 114.

[55] The Butler Inquiry commented on this by recommending that "arrangements must be made for the future that avoid putting the JIC and its chairman in a similar position.

[56] See Richard J. Aldrich, "Regulation by Revelation? Intelligence, the Media and Transparency," in *Spinning Intelligence: Why Intelligence Needs the Media, Why the Media Needs Intelligence*, ed. Robert Dover and Michael S. Goodman (New York: Columbia University Press, 2009).

[57] Winfred Göpfert, "The Strength of PR and Weakness of Science Journalism," in *Journalism, Science and Society: Science Communication between News and Public Relations*, ed. Martin W. Bauer and Massimiano Bucchi (New York: Routledge, 2007), p. 215.

intelligence—has been subjected to a transformed media logic. This media logic has tended to exaggerate criticism of decision makers beyond the point that would be reasonable from the idealistic perception of a more confined watchdog role. Media is, according to some interpretations, simply exploiting inaccurate criticism on the pretext of making a social contribution, while in fact it is making a commercial one.[58] But at the same time the production process of major media has moved in an opposite direction, with fewer resources devoted to independent scientific journalism and thus an increasing reliance on material produced by the scientists themselves, their media communicators, or their public relations experts.[59] The media process has remained focused on the news angle to be the fastest, newest, and most fantastic, perhaps adding the most scandalous to the list. This has resulted in a dual pattern, where science on the one hand is subjected to hype, and on the other subjected to recurring media gauntlets.[60]

Another fundamental change is the increasing employment of media by the scientific community in lobbying for funding, a process that gathered momentum in the United States in the 1990s as research funds declined. "Selling science," however, is a complex process full of contradictions. One such complicating factor has been the reluctance of scientists to interact with media, based on a widespread assumption that explaining scientific problems to journalists is a waste of time, since the subject matter normally is too complicated and the journalists will get it wrong anyway.[61] Another complication is the blurred distinction between information and promotion, between the role of the scientific community to inform the public and the role of promoting vested interest by using media and the public as pressure groups. These can be described as "the science lobby," defined as a group of people using the media to forward a set of over-arching goals to secure more funding while retaining control of the funds and choices of projects, methods, and procedures and limiting demands from society to those goals that the scientific community can meet with reasonable success.[62] Dorothy Nelkin describes the combined

---

[58] Björn Fjaestad, "Why Journalists Report Science as They Do," in *Journalism, Science and Society: Science Communication between News and Public Relations*, ed. Martin W. Bauer and Massimiano Bucchi (New York: Routledge, 2007). For scientific promotion in media, see Wilhelm Agrell, "Selling Big Science: Perceptions of Prospects and Risks in the Public Case for the ESS in Lund," in *In Pursuit of a Promise: Perspectives on the Political Process to Establish the European Spallation Source (ESS) in Lund, Sweden*, ed. Olof Hallonsten (Lund: Arkiv 2012).

[59] Göpfert, "The Strength of PR and Weakness of Science Journalism."

[60] Dorothy Nelkin, *Selling Science: How the Press Covers Science and Technology* (New York: Freeman, 1995).

[61] Ibid., pp. 7–8.

[62] Fjaestad, "Why Journalists Report Science as They Do," p. 124.

effect of scientific "mystique" and increased pressure to employ media for strategic promotion. In the case of Interferon, the protein presented as a breakthrough in cancer treatment in the early 1980s, the scientists, far from being a neutral source of information, actively promoted Interferon and thus shaped the flawed media coverage.[63] When Interferon failed to deliver the promised results, the whole campaign backfired into a media-rich scientific scandal.

Nobel Prize winner Kenneth Wilson summarized the key elements in selling Big Science in a statement after successfully convincing the National Science Foundation to support the supercomputer program in the mid-1990s. His arguments were not scientific but political and ideological: without the program the United States would lose its lead in supercomputer technology. Wilson was amazed at how the media picked up on what he described as "a little insignificant groups of words" and he concluded: "The substance of it all [supercomputer research] is too complicated to get across—it's the image that is important. The image of this computer program as the key to our technological leadership is what drives the interplay between people like ourselves and the media and force a reaction from Congressmen."[64]

A simple and idealistic linear model of science communication seems to help very little for the understanding of these inter-linked transformations in science coverage in the media, scientific lobbying, and the shaping of public perceptions of science as a whole and of specific projects in particular. Indeed, the literature on the subject, the scientific/media interface, seems to revolve around four main themes:

- The assumed "unexplainable" nature of advanced scientific projects and activities.
- Media as a channel for scientific lobbying and public relations.
- The power of catchwords and compelling nonscientific arguments in the framing of usefulness and impact.
- The lingering threat from the—perhaps over-active—media watchdogs.

Turning to intelligence we can observe a corresponding pattern, although with some very different premises and a delay in timing. If scientific journalism grew increasingly critical from the 1970s onward, this was not the case for intelligence, even if the large-scale public scandals of the 1970s, both in the United States and several other Western countries, unleashed a wave of media coverage that was both unwanted and unflattering about intelligence

---

[63] Nelkin, *Selling Science*, p. 6.
[64] Ibid., p. 125, quoted from the *New York Times,* March 16, 1985.

operations and their doubtful ethics and legality. But the change after 2003 has been a more profound one, focusing not only on surveillance, citizen rights, and ethics of certain operations but on the core of the intelligence process—the quality, trustworthiness, and relevance of the intelligence flow and subsequent assessments. Intelligence has not only come in from the cold but has entered the heat of media scrutiny, in many ways far more ill-prepared than contemporary scientists. With this said, the four themes listed earlier have their obvious equivalence in intelligence:

- Intelligence is not only unexplainable to laypeople but explaining it is an impossible task due to secrecy, as highlighted in Tony Blair's foreword to the September Dossier.
- Media proved to be a powerful channel for dissemination and marketing—but with an unforeseen powerful potential for backlash.
- Intelligence lingua franca is not understood, but catchwords are. The "45-minute warning" in the September Dossier is possibly the most powerful marketing remix of an intelligence assessment ever. In this sense Alastair Campbell was right and those feeling unease in Defence Intelligence were wrong.
- As intelligence was to experience, the watchdog instinct and media logic were grossly underestimated. For media, intelligence simply delivered two successive news items: first the WMD story and then the blockbuster of the intelligence scandal of the missing WMD.

In these circumstances, is politicization the problem, or at least part of the answer? In Robert Gates's 1992 warning speech, politicization was framed as the overriding menace, a forbidden fruit that could undermine the professional foundation of intelligence, depriving it of key values and turning intelligence into an anti-thesis, the production of biased knowledge and distortion of the truth. The disaster in 2002–2003 seemed to confirm this perspective; intelligence was politicized and its knowledge claims flawed. The proud banners had been torn, and the long-term repercussions were still visible after a decade in the failed efforts to raise domestic and international support for action against the Syrian regime after the August 2013 chemical weapons attack. A closer reading, however, reveals a more complex picture of events in 2002–2003. Maybe intelligence was not as disastrously wrong as hindsight indicated. Maybe other related factors played a role, such as the wish to be perceived as relevant, called for, read, and quoted. Maybe intelligence, at the end of the day, had not mattered that much at all, and anyway not in the classical sense of supplying information for action. And maybe politicization is not the dark threat to truth, but an epistemic drift in *all* socially relevant knowledge production, another word to describe the realities in the world beyond the protection of the walls of secrecy and the ivory tower.

Nevertheless, politicization has an unavoidable negative connotation, both in intelligence and in science. While science could—and in some respects should—be conducted for the sake of knowledge production in its own right, intelligence is by definition far more user- and usefulness-oriented. True, there is, at least in theory, a similar need for basic research in intelligence, but the pressure of immediate demands seldom allows for this, and the realities of bureaucratic politics would not allow such safe havens to survive for long. And the dynamics of knowledge production do not seem to press for a return to the Ivory Tower in hardly any domain. The more "intelligence-like" research problems become, the more they will inevitably be subjected to the cross pressures where politicization could emerge. However, it is hard to see a corresponding change in intelligence becoming more scientific in methods and approaches, and less subjected to politicization. There is no calm backwater left, and all knowledge production with any relevance or potential impact is swept out onto the vast arena of media-driven politics, culture, and social interaction. Out there, they will all intermingle, the politicians, the stakeholders, the researchers, the intelligence officials, and the journalists, all struggling with handling uncertainty and confronted with the recurring question: "What's new?"

9

Beyond the Divide

Framing and Assessing Risks under Prevailing Uncertainty

On December 3, 2010, the Norwegian Security Service (PST) received a list from the Customs Directorate with names of 41 people who had purchased products on the Internet from the Polish firm *Likurg*, which specialized in chemical substances and pyrotechnics. The list received little attention; no check was made against the security service files or other sources in subsequent months, and the list was still awaiting processing when a twin terrorist attack struck the Oslo area on July 22, 2011. Only afterward was it discovered that one of the names on the list was that of Anders Behring Breivik, the arrested perpetrator of the attacks that claimed 77 lives in a massive car bombing of government headquarters in central Oslo and a subsequent systematic massacre of defenseless teenagers at a political youth camp on the tiny island of Utøya. The Likurg list thereby was predestined to attain the same status and provoke the same contrafactual disputes as the pre-911 FBI Phoenix Memo and the pre-Pearl Harbor Japanese Wind Warning cable.

Revisiting the Unknown Unknown

In brief, the backstory to the lack of warning was the following: in August 2010 the Norwegian Customs Directorate was, along with a large number of customs services throughout the world, invited to participate in a project named Global Shield, initiated by the US Department of Homeland Security through the World Customs Organization (WCO).[1] The aim of the project was

---

[1] The terrorist attacks and the efforts of the Norwegian authorities to prevent and handle the consequences were investigated by a government commission, the *22 July Commission*, presenting its findings in August 2012. The main report is in Norwegian, but there is an abridged summary in English. References here are to the main report NOU 2012:14, available at http://www.22julikommisjonen.no/. The 22 July Commission, "Findings of the 22 July Commission " (Oslo, 2012).

to initiate a global effort to monitor and control the transfers of substances and components that could be used for constructing improvised explosive devices (IEDs), a threat that was spreading from the conflict areas in the Middle East and Central Asia. Norway was far out on the periphery but had nevertheless suffered losses to IED attacks against its International Security Assistance Force (ISAF) troops in Afghanistan.

The invitation to take part in Global Shield was also forwarded to the PST, but here the invitation was mistaken as merely a piece of general information from colleagues in the customs service. The officer at the Customs Directorate assigned to the project quickly realized that a full check of the substances on the Global Shield list would be impossible, due to the large quantities of fertilizers handled by the leading Norwegian company. Having only limited time to devote to the project, the officer, still wanting to make a contribution to proactive counterterrorism, looked for something more manageable to investigate. One way of limiting the data was to exclude the companies and look at private imports alone, based on the presumption that it was here that suspicious transaction could be found. Private imports were also relatively easy to separate from the rest. A first list, containing eight names, was found and sent to the PST in October. This first batch of names was checked, but no matches were found with names in the Security Service files.[2]

At the Customs Directorate the officer in charge continued his efforts. In the subsequently available information in the Global Shield database he found that a Norwegian citizen had imported some of the monitored substances from what appeared to be a private person in Poland. Looking closer into this, the officer discovered that this in fact was a company named Likurg, with a homepage in Polish, English, and Norwegian. Suspecting that there could be more transactions, he checked against the Norwegian currency register and found 63 transactions by 41 individuals. This was the Likurg list forwarded to the PST in early December 2010. The list only contained information on the currency transactions, not the nature of the purchases. Anders Behring Breivik's transaction was the smallest amount on the list, 122 Norwegian crowns (equal to less than 20 USD). Breivik had started his systematic purchases of bomb components and associated equipment in October 2010, and it was a pure coincidence that his single purchase from Likurg made in November happened to coincide with the customs officer's check in the currency registry. The item itself, however, was a critical component—the fuse that eventually was used to detonate the 950 kg car bomb on July 22, 2011.

After the first check in October, the PST did not take any further action, and the subsequent discovery of the Likurg transaction provoked no response.

---

[2] Ibid., p. 374.

The reason was a combination of heavy workload and the internal inability to decide whether the case belonged to the counter-proliferation branch or the counterterrorism branch of the security service. The case was finally given to an arms and weapons intelligence analyst at the counterterrorism branch in late April 2011, but by that time the analyst was due for a long leave. When the bomb exploded, the documents from December 2010 were still unprocessed.[3]

After the attack, the PST was faced with the awkward question of assessing the significance of the unprocessed list. With hindsight it was inevitable that the potential significance of the December 3 list, with the future perpetrator's name on it, should be a crucial point in a postmortem. Would it have been possible to prevent the successful preparation and subsequent execution of the attacks with a more vigilant stance, if available leads had been pursued and the right questions posed? In a self-evaluation by the PST, published in March 2012, the service devoted considerable effort to show that even with a timely analysis of the material from Global Shield the preparations under way by Breivik would not have been uncovered. He had no Security Service file, no significant criminal record, and no deviating activities on the Internet. Besides that, the Security Service was prevented from taking any action against an individual if there were no specific hypothesis that motivated such a move.[4]

Would a more scientific approach to this specific intelligence challenge have made any difference? And would scientists have been more capable of maneuvering through the fog of uncertainty surrounding weak signals? No doubt the PST, like so many other intelligence and security services in similar predicaments, was simply unfortunate in getting tangled up in the all-too-well-known interaction of information overload, compartmentalization, and legal caveats. After the attack it was left fending off criticism, using the familiar lines that the list would not have mattered anyway, given the standard operating procedures and things being as they were. But a scientific analysis would have been faced with the same kinds of blind spots and barriers; the same friction between disciplines, institutions, and research groups; the same prevalence of dominating paradigms resulting in the same uphill battle for anyone trying to pursue a lead that by definition was a dead end, provided that the entire set of established theories was not invalid and that the impossible in fact was the most probable

[3] Ibid., p. 378.

[4] The PST (Norwegian Security Service) writes: "3 December 2011 there was, according to our assessment, no valid ground for initiating a full entry on Anders Behring Breivik and the other 40 persons on the Customs Directorate list. There was no working hypothesis to relate such an entry to, and the entry in itself would not have met the standards for relevance and purpose" (our translation). *PST Evalueringsrapport* (PST Evaluation Report) (Oslo: PST, March 16, 2012), p. 23.

explanation. If anything deviated from the expected pattern, it is not the failure of PST to follow an odd lead but rather the fact that this odd lead was discovered at all. The search was inventive—and in some respects possibly legally
doubtful—independently looking for something the Customs official could
not imagine, driven as he later explained to the commission by a combination
of duty and curiosity.

As we discussed in chapter 3, the positivist "Jominian" approach is to
eliminate uncertainty and to find the right answer, while a "Clausewitzian"
approach accepts uncertainty as the basic precondition, be it in war, intelligence assessment, or scientific inference. The 122 Norwegian crown purchase
on the Likurg list was from the Jominian perspective insignificant information without any meaningful bearing on the prevailing threat paradigm and
the body of information and assessments that constituted it. The major error
was perhaps not overlooking the list or even failing to discover the systematic preparations for the attack, but the lack of any appreciation that ambiguous pieces of critical information were bound to be overlooked, and that the
systematic efforts of the intelligence machinery to eliminate uncertainty in
fact transformed it into a mode virtually primed for disaster. Ignorance, as
Mark Phytian notes, is different from uncertainty, where we know that there
is something we have incomplete or inaccurate knowledge of. Ignorance is
the unknown unknown, the fact of not knowing what we don't know. It is
boundless and virtually leaves us in the dark.[5]

Every knowledge-producing system faced with assessments of probabilities and risk, whether in intelligence, civil contingency, cross-disciplinary
research, or investigative journalism, is likely to fail from time to time when
faced with deviation from known or expected patterns. But for the Likurg
list, things looked different. The only possible outcome was bound to be
refutation through lack of confirmation. The system had, as the PST, possibly unintentionally, disclosed in the self-evaluation, effectively closed every
door for another outcome: only a successful terrorist attack would alter the
track record–based assessment that domestic right-wing extremism did not
pose a threat to the state and its institutions.[6] Crucial to the fate of the Likurg
list was thus not the inability to take action or the methods to analyze and
draw inferences from the available information, but something far more fundamental—the lack of a hypothesis. The list was an answer without a question, and what was worse, without even a remote possibility of framing such

[5] Phythian, "Policing Uncertainty: Intelligence, Security and Risk," *Intelligence and National Security* 27, no. 2 (2012): 187–205.

[6] On PST threat assessments from 2001 to 2011, see Chapter 4 of The 22 July Commission, "Findings of the 22 July Commission,"

a question. Facing the unthinkable, the unknown unkown, was in this case simply unmanageable.[7]

As Robert Jervis found in his internal postmortem on the failed predictions of the Shah's downfall 1978, the CIA analysts *thought* they were operating with a testable hypothesis on regime stability, while in fact the only way to refute the assessment of stability was the actual fall of the regime, a contingency lying outside the prevailing paradigm. Faced with these methods to combat uncertainty, any black swan is bound to be not only undetected but also virtually undetectable. No perpetrator or hostile actor is safer than the one operating in this dead zone of intelligence, where only bad luck in the form of the upside of Clausewitzan friction could change the otherwise inevitable march toward disaster for the victim, a situation that by definition is characterized by the sudden appearance a of low probability–high impact event and the breakdown of cognitive structures and frames of reference.[8] Knowledge here tends to become a bubble, a body of more or less solid information and threat assessments, but one surrounded by an undefined entity of ignorance, of the unknown unknown, or perhaps even unknowable unknown. Without entering into the contrafactual blame game so tempting in cases like this, it is hard to avoid the conclusion that the main fallacy prior to the July 22 attacks was not the lack of adequate information or the limits of intelligence estimates, but something far more fundamental: the invisibility of a potentially disastrous ignorance of emerging threats as such.

Cracking the Paradigm: Activity-Based Intelligence

The Likurg example drives home the question of whether innovations in approaching intelligence and intelligence analysis might offer some hope if not of locating the unknown unknowns, then at least in diminishing the fog obscuring the fact that critical questions are not only going unanswered but also unasked, and that their answers may appear before the question is

---

[7] For a further discussion of intelligence aspects, see Wilhelm Agrell, "The Black Swan and Its Opponents. Early Warning Aspects of the Norway Attacks on July 22, 2011" (Stockholm: National Defence College, Center for Asymetric Threat Studies, 2013).

[8] For the chaotic nature of crisis, see Arjen Boin et al., *The Politics of Crisis Management: Public Leadership under Pressure* (Cambridge: Cambridge University Press, 2005), pp. 2 ff. Crises can be defined as largely improbable events with exceptionally negative consequences; see Tom Christensen, Per Lægreid, and Lise H. Rykkja, "How to Cope with a Terrorist Attack?—a Challenge for Political and Administrative Leadership," in *COCOPS Working Paper No 6* (2012), p. 4, available at http://www.cocops.eu/publications/working-papers.

formulated. What is now called "activity-based intelligence" developed in response to the changed threat that we have emphasized so often in this book.[9] The quotes cited earlier capture the change: imagery used to know what it was looking for and be looking for things; now it doesn't know what it is looking for and it isn't looking for things. Harold Brown's line about the Soviet-American nuclear competition—"when we build, the Soviets build; when we stop, they build"—conveys the sense of how unconnected the two adversaries were; compare this to terrorists, who are constantly looking for both vulnerabilities to attack and places to hide. The targets of the fight against terror are small, not big; adaptive, not ponderous; and, perhaps most important, lack both signature and doctrine. A Soviet T-72 tank had a distinctive signature, one easily captured by a single intelligence discipline, imagery, and was embedded in doctrine: if one was seen, a given number of its brothers could be expected as well. The intelligence process was linear and inductive, reasoning from a single sighting to a larger category of phenomena.

The terrorist target, and especially the challenge of finding the "bad guys," was utterly different. No predictable or semi-permanent signature distinguished the would-be implanters of IEDs from ordinary pious Muslims at mosques. Even if a cellphone signature could be resolved to a suspected terrorist, the next day could easily see a different phone entirely. Activity-based intelligence (ABI) as it developed was especially impressive in the war zones of Iraq or Afghanistan in uncovering would-be terrorists and their networks. The US Pentagon defined ABI as a discipline of intelligence where the analysis and subsequent collection is focused on the activity and transactions associated with an entity, population, or area of interest.[10] The sequence of events that led to the raid on Osama bin-Ladin's compound is a case in point. It remains to be seen how relevant ABI techniques will be beyond manhunting in the fight against terror. But for our purposes, the question is whether the basic tenets of ABI can be provocative in opening up some of the rigidities in intelligence—and especially intelligence analysis—that have run through this book. Do they suggest another paradigm? And is that paradigm more scientific—or less?

As outlined in Chapter 3, the canonical "intelligence cycle" has been criticized, including by us, but usually on the basis that it is too stylized a representation of a much messier reality of skipped steps and simultaneous actions. But a comparison with ABI underscores how much the "cycle" enshrines a whole series of assumptions that are unhelpful and unwise. For starters, it assumes

---

[9] For a nice primer on activity-based intelligence, see Kristin Quinn, "A Better Toolbox," *TrajectoryMagazine.com* 2012.

[10] "Activity Based Intelligence Knowledge Management, USD(I) Concept Paper," (Washington: Undersecretary of Defense, Intelligence Draft, 2011).

that we know what we're looking for. The process begins with "requirements." By contrast, ABI assumes we don't know what we're looking for. The cycle also assumes linearity. Sure, steps get skipped, but the basic model is linear. Start with requirements, then collect, then process, then analyze, and ultimately disseminate. That sequence also assumes that what is disseminated is a "product"; it tends to turn intelligence into a commodity. Even the best writing on intelligence, like Roberta Wohlstetter's book on World War II through the wonderfully written 9/11 report, is linear: "dots" were there, if only they had been connected. But think about how non-linear most of thought is, and surely most of creativity. This is what, in ABI, is "sequence neutrality." In intelligence, as in life, we often may solve a puzzle before we realize the puzzle was in our minds. Or to return to the analogy with medicine, notice how many drugs were discovered by accident to be effective for one condition when they were being used for something else entirely. The discovery was an accident, hardly the result of a linear production function. Or as in the Likurg case, that list may appear before the puzzle to which it is the answer is—or perhaps can be—framed.

Perhaps most important and least helpful, traditional intelligence and the cycle give pride of place to *collection*. Collection drives the entire process: there is nothing to analyze if nothing is collected against the requirements that have been identified. When analysis or postmortems reveal an intelligence "gap," the first response is to collect more to fill that gap. But the world is full of data. Any "collection" is bound to be very selective and very partial. Perhaps slightly paradoxically, ABI reminds us that data are not really the problem. Data are everywhere, and getting more so. China has, by one estimate, one camera for every 43 people. Data are ubiquitous, whether we like it or not. Privacy, as it was traditionally conceived, is gone, or going, again for better or worse.

And not only does intelligence privilege collection, it still privileges information *it* collects, often from its special or secret sources. As noted earlier, the founding fathers of intelligence in the United States (alas, there were few mothers aside from Roberta Wohlstetter) worried, as Kent put it, that the concentration on clandestine collection would deform open collection. And so it has, in spades. For ABI, by contrast, all data are neutral. There are no "reliable sources"—or questionable ones. Data only become "good" or "bad" when used. This neutrality of data is, not surprisingly, perhaps the most uncomfortable challenge of ABI to the traditional paradigm of intelligence, for so much of the cycle has been built around precisely the opposite—evaluating information sources for their reliability. ABI here rather resembles post-modern social science and humanities dealing with discourse analysis, perceptions, and symbols.

At this point, ABI intersects with "big data." So much of traditional intelligence analysis has been very limited in information, assuming that only "exquisite sources" matter. It has developed neither the method nor the practice of looking for and dealing with large amounts of data, but often ones that are tangential, or partial or biased in unknown ways. There is lots of information out there that doesn't get used. The promise of "big data" may turn out to be the analytic equivalent of the technology bubble. But the proposition that there is lots of it out there to be incorporated is apt. So is the admonition to collectors and analysts that things they see now, without relevance to their current concerns, may be important in the future. To be sure, that admonition not only requires an enormous change in culture from the view that only information relevant to my hypothesis matters, but it also imposes a huge knowledge management challenge. And that, too, is a piece of breaking the paradigm, for traditional intelligence analysts mostly didn't conceive of themselves in the knowledge management business. Imagine a bumper sticker: "big data can empower non-linear analytics."

Two other elements of ABI are suggestive in opening the paradigm. ABI turns on correlation: looking at patterns of activity to determine pattern of life and so networks. It is agnostic about causation, looking first for how things fit together. But the idea of correlation also opens up traditional analysis, for the linear process can all too easily lead to linear thinking—"drivers" impute causation. But causation in human affairs is a very complicated affair. Imagine if analysts grew up in a world of correlations—indicators—not drivers that imputed causation, where none usually obtains. They might be attuned to *connections*, perhaps even ones not natural to analysis, or to life. The idea that a busker in Tunisia who set himself on fire might in turn propel activists in Egypt to action surely wouldn't occur but might be less easily dismissed out of hand.

Activity-based intelligence is also inherently collaborative, a challenge to the practice of intelligence analysis if not necessarily to the paradigm. It is distinguished from traditional pattern analysis mostly in that while patterns could be uncovered by information from several intelligence sources, or INTs, most of the time a single INT was good enough: recall those Soviet T-72 tanks whose signature was visible to imagery. But multi-INT or ABI has required analysts to work closely together, often in real time. It has made them realize not only that other disciplines can contribute to solving *their* problem but educated them in how those disciplines work. In a few cells, analysts from different disciplines, often separated geographically, have employed new tools of social media, like chat rooms and Intellipedia, to exchange information in real time and build a picture of what is happening around, for instance, the launch of a North Korean missile.

So imagine a "might have been" and a "might be." With regard to the Arab spring, it was hardly a secret that the countries involved were unstable. A generation of work by the CIA, other services, and their academic counterparts had created indicators of political instabilities, which were then turned into countries to watch. Those lists tended to be noted by policy officers but were never found very helpful because they could not specify when or how explosions might happen. The point was a fair one, for from the perspective of traditional intelligence, the idea of a street vendor setting himself on fire in Tunisia and the action leading to uprising in a half dozen other countries was a black swan indeed. Had anyone suggested it, the suggestion would have fallen in the "outrageous" category and been dismissed as such. It would not, perhaps could not, have been recognized as the answer to yet another puzzle not yet framed.[11]

But what if? What if analysts had been looking for correlations, not causation. Suppose they had been monitoring Twitter or other open sources, and laying those against other sources both open and secret. The process could hardly have been definitive. But it might have provided hints, or gotten analysts to suggest other things to look at. At a minimum, the process might have helped to weaken the power of mindset—in this case, for instance, that nothing in Tunisia could influence what happened in Egypt. It might have been a way to weaken the grip of mindset, by keeping the focus on data.

In the category of "might be," consider China's future. It has long been assumed that, sooner or later, growth would slow and that the social change driven by economic growth would lead to political change.[12] Now, there is a good argument, though hardly proof, that the timing is "sooner." Demographics is one powerful factor. China will soon fall off the demographic cliff, as the fruits of the one-child policy mean a rapidly aging population. And critically, China will begin to age when it is still relatively poor, and thus growing old for the Chinese will is not be like aging in Japan or Europe.

Moreover, for both economic growth and political change, the key indicator is for China to reach a per capita income of $17,000 per year in 2005 purchasing power parity.[13] That will occur in 2015 if China continues to grow at 9–10 percent annually, in 2017 if growth slows to 7 percent. Again, correlations are not predictions, but no non-oil-rich country with a per capita income over that level is ranked as other than "free" or "partly free" by Freedom House, while China is now deep in the "non-free" category.[14] As China has gotten richer, its people have become less inhibited in raising their voices, and the Internet provides them

[11] For the shortcomings of a traditional intelligence approach in the face of social changes like the Arab spring, see Eyal Pascovich, "Intelligence Assessment Regarding Social Developments: The Israeli Experience," *International Journal of Intelligence and Counterintelligence*, 26, no. 1 (2013): 84–114.

[12] This section draws on Gregory F. Treverton, "Assessing Threats to the U.S.," in *Great Decisions 2013* (New York: Foreign Policy Association, 2012), pp. 101ff.

[13] Henry S. Rowen, "Will China's High Speed Ride Get Derailed?," unpublished paper, 2011.

[14] Henry S. Rowen, "When Will the Chinese People Be Free?," *Journal of Democracy* 18, no. 3 (2007): 38–52.

increasing opportunities, notwithstanding the government's zeal in censoring it. The government has stopped enumerating what it calls "mass incidents"— strikes or protests usually carefully crafted to stick to local issues—but an estimate for 2010 puts these at 160,000, up from 10,000 in 1995. Another indicator is the visible nervousness of China's leaders about the polity over which they rule.

On the economic side, one recent study found that high growth in non-oil exporting countries ended at a per capita income of about $17,000.[15] There are good economic arguments for the threshold: the payoff for shifting workers from agriculture to industry diminishes, as does the benefit from adapting foreign-developed technologies, and an undervalued exchange rate also is a factor. In this study, growth slowed from 5.6 to 2.1 percent per year. A drop of that magnitude in China's growth would take it to the 6–7 percent range. Again, there is nothing magic about correlation, and China might escape the threshold by developing its own internal undeveloped "country" in western China. Moreover, much of the world would be delighted with 6 percent growth.

Yet economics and politics will intersect. Both Chinese and foreign analysts have long understood the major "fault lines" ("adversities") that might seriously affect China's ability to sustain rapid economic growth.[16] How might these adversities occur, and by how much would they affect China's growth? If growth slowed, what sectors and what people would be hit? How much would unemployment and underemployment increase, including for college graduates, which already is a problem? The available data are plentiful, from government numbers of protests to middle-class behavior. If those were continually laid against other sources, like Chinese bloggery and whatever hints were available from human intelligence or political reporting, again, the likelihood of definitive warning would be very low. But the process might suggest new places to look and open analysis to unsuspected correlations.

Uncertainty versus Urgency

National security intelligence during the 20th century essentially combined a long-range enterprise of strategic intelligence with current intelligence, which was often repetitive and monotonous. The saying that Cold War intelligence was characterized by years of boredom and seconds of terror reflected this

---

[15] Barry Eichengreen, Donghyun Park, and Kwanho Shin, "When Fast Growing Economies Slow Down: International Evidence and Implications for China," *National Bureau of Economic Research Working Paper* 16919 (2011), available at http://www.nber.org/papers/w16919.

[16] Charles Wolf et al., *Fault Lines in China's Economic Terrain*, MR–1686–NA/SRF ed. (Santa Monica: RAND Corporation, 2003).

dualism. Strategic intelligence, the ongoing work in the West on the giant puzzles of Soviet capacity, was driven by requirements to gradually reduce a recurring uncertainty. Current intelligence during the years of boredom validated the puzzle, and in the seconds of terror forced urgent reassessment and crisis management. In the major armed conflicts, urgency dominated. Policymakers and headquarters and field commanders often had to take rapid decisions based on incomplete or ambiguous intelligence, for better or worse relying on their own judgment. The fact that intelligence veterans tended to describe Cold War intelligence as dull and over-bureaucratized compared to World War II did not just reflect a preference for glorifying past times. It was a fact.

The war on terror, one affecting virtually every country and in some respects every individual, altered the entire setting for national security intelligence. The puzzle, if there ever had been one, was shattered and infeasible to even imagine. Without a stable over-arching threat, puzzles could simply be laid and re-laid in an endless process that nevertheless did not reflect the complexity and interactive character of the threat.

This profound change from puzzles and validated threat assessments to the realm of complexities was heralded by the rise of the international terrorist threat. But post-9/11 terrorism was not the sole cause, only the most visible aspect of a wider undercurrent, that forced intelligence analysts, technical and scientific expertise, and national leaders into a new terrain of crisis decision making characterized by an immense real-time flow of information, vastly complex issues, and a subsequent urgency to both comprehend and react on rapidly unfolding events. How could knowledge production and expert advice be reshaped and framed to function under such contingencies? Or is that task simply impossible, with intelligence becoming instead a means of trying to keep up confidence and legitimize the decisions or non-decisions taken for other reasons in the public arena?

There was no specific emergency pressing the Norwegian Security Service prior to the July 22 attacks, only the all-too-well-known volume of routine work and ongoing criminal investigations making any new input unwanted and unwelcome.[17] The circumstances were in this respect entirely different when the WHO and national health authorities over the world were faced with the renewed outbreak and rapid spread of the swine-flu (H1N1) in the spring 2009. In the first outbreak in 1976 the leading scientists overstated the

---

[17] An external review of the Norwegian Security Service (PST), led by Ambassador Kim Traavik, concluded that the service to a high degree had concentrated on ongoing matters and lacked attention for things outside this sphere. The PST had, in the view of the investigation, been unable to see the broader picture and how it transformed. "Ekstern Gjennomgang Av Politiets Sikkerhetstjeneste. Rapport Fra Traavikutvalget" (Oslo: Justis– Og Beredskapsdepartementet, Dec. 3, 2012).

probability of a pandemic outbreak, knowing that nothing short of a real possibility would compel the policymakers in the Ford administration to take action. As it turned out, there were only a few cases of the flu and no spread at all, and the vaccination program well under way was discontinued. In 2009 the rapid spread and thus the prospect of an emerging H1N1 pandemic preceded the decision to start vaccine production and mass vaccination. The threat was a reality when the decision to react was taken, but the tight time frame available for preventive vaccination meant that some known uncertainties on side effects remained.

From October 2009 to March 2010, approximately 60 percent of the Swedish population was vaccinated again H1N1 following the decision of the WHO to declare the disease pandemic. During spring and summer 2010 an unexpected large number of cases of narcolepsy among children and teenagers were reported both in Sweden and in Finland, and the increase was suspected to be linked to the specific type of flu vaccine that had been used. Large-scale comparative cohort studies conducted by the Swedish Medical Product Agency concluded in March 2013 that there was a correlation and that the risk of developing narcolepsy were three times higher for inoculated persons under the age of 20, and twice as high for persons between the age of 20 and 30, compared to the unvaccinated population.[18] As the N1H1 infection turned out to be far less serious than expected, the inevitable question was whether vaccination in this case had done more harm than good.

The medical experts, however, defended the decision, referring to the uncertainty at an early stage about the seriousness of the infection, and the limited time available for conducting tests on possible negative side effects of the vaccine. Urgency was traded for uncertainty—or rather one uncertainty for another. The problem here was rather one of insufficient communicated uncertainty and thereby understated risk. In order to facilitate a near-complete vaccination—the prerequisite for halting a pandemic spread—the public confidence in the effectiveness and safety of the vaccine had to be maintained. The main argument from the authorities was one of negative verification. Tests of the vaccine during the short time span available between development and delivery of the first batches did not indicate any statistically significant side effects, but the inescapable lack of knowledge of any later effects was not clearly underlined.

---

[18] "Registerstudie med focus på neurologiska och immunrelaterade sjukdomar efter vaccination med Pandemrix,Medical Production Agency, Uppsala, March 26, 2013." The study incidentally did not show any increase in the kinds of side effects reported from the 1976 US mass vaccination.

As Richard Betts notes, criticism of the US intelligence estimates on the Iraqi WMD programs on the ground that they turned out to be wrong is fundamentally flawed. Getting it right is not the minimum standard for good assessments, or as Betts writes: "Hindsight inevitably makes most people assume that the only acceptable analysis is one that gives the right answer."[19] Poor assessment methods and hunches can, as in the case of the Cuban missile crisis, turn out to be correct. But should we turn to astrology if an astrologist happens to hit the mark? Or advise analysts to follow every interesting lead in the hope that they, when venturing out into the unknown unknown, just might stumble over something out there? Activity-based intelligence could certainly help open the mental box, but it could also transform that box into another endless frontier. In cases of deep uncertainty, neither scientists nor intelligence analysts can be expected to be dead right; they can only be expected to reach the best possible conclusions under the prevailing circumstances, which from time to time is not very satisfactory. Perhaps it is here, in the framing and sense-making of these prevailing circumstances, that the main challenge lies, and where sins are most frequently committed, especially creating and spreading illusions of certainty in domains where no absolute certainty is possible. That sin is accompanied by others—the reluctance to recognize uncertainty, the inadequate efforts to describe and quantify it, and the flawed assumption that somehow those relying on the assessments would be better off without knowing too much about it. Like the diner tasting sausage, perhaps they don't want to know too much about how it was made.

The concept of *risk society*, coined by German sociologist Ulrich Beck in 1986, highlights the growing uncertainty surrounding the gains of modernization and the negative fallout in terms of environmental impact and the increasing risks from complex technologies and vulnerable social systems. In the risk society, science and technology are generators of unintentional risk, but at the same time they are also the tools to discover, warn of, and avert risks. In the risk society it is the management of risks and subsequent trade-offs and conflicts that transcends both politics and everyday life.[20] The risk society has been regarded by some as an exaggerated exponent of the technology skepticism and environmental alarmism of the 1970s and 1980s. Yet it is hard to

---

[19] Betts, *Enemies of Intelligence: Knowledge and Power in American National Security* (New York: Columbia University Press, 2007), pp. 122 ff.

[20] Ulrich Beck, *Risk Society: Towards a New Modernity* (London: Sage, 1992). The book was originally published in German with the title *Risikogesellschaft: auf dem Weg in eine andere Moderne* (Frankfurt: Suhrkamp, 1986). For a discussion of the impact of Beck's concept, see Joris Hogenboom, Arthur P. J. Mol, and Gert Spaargen, "Dealing with Environmental Risk in Reflexive Modernity," in *Risk in the Modern Age: Social Theory, Science and Environmental Decision-Making*, ed. Maurie J. Cohen (Basingstoke: Palgrave, 2001).

avoid the conclusion that the concept nearly three decades later has become a reality in terms of public perceptions and policy priorities. We have gone from external and internal threats to national security to a broad risk matrix of societal risks with varying assumed probabilities and impacts. The cases of intelligence and scientific assessments we have dealt with in this book gained much of their significance in relation to this emerging awareness of risk management as one of the key roles of society.

The Nobel laureates at the 2012 round table were, as discussed in the introduction, obviously very much aware of this, and of the credibility gap looming between the public in their expectations and the experts on the nature and perceptions of uncertainty. The experts could be tempted to oversell assurances, which happened with the Fukushima disaster, as summarized by medicine laureate Shinya Yamanaka: all the experts kept saying nuclear energy was safe, but it turned out this was not the case. The disaster not only struck the nuclear power plant and the society in the contaminated area, but it also destroyed faith in scientific competence, impartiality, and ability of the experts to manage risks. Scientific expertise time and again faces the awkward dilemma of balancing expectations and uncertainty, and the disastrous consequences of loss of trust among the public. Transparency was, in Yamanaka's opinion, the answer to the credibility loss caused by the Fukushima meltdown; the experts had to be as honest and open as possible and deliver their message in words that could be understood by laypeople. Intelligence veteran Michael Herman's earlier quoted commented described the public fallout of the Iraqi WMD intelligence disaster in a similar way, underlining the element of expert deception and loss of public trust.

Turning intelligence into science, while certainly useful to some extent and in some instances, nevertheless misses the point that several of the fundamental challenges are neither specific to intelligence nor more successfully solved in the scientific domain. The essence of intelligence is hardly any longer the collection, analysis, and dissemination of secret information, but rather the management of uncertainty in areas critical for security goals for societies. This is also a definition valid for increasingly important fields of both applied and urgent basic research created by the emerging risk society. We are living in a social environment transcended by growing security and intelligence challenges, while at the same time the traditional narrow intelligence concept is becoming increasingly insufficient for coping with diffuse, complex, and transforming threats. Looking ahead, the two 20th-century monoliths, the scientific and the intelligence estates, are simply becoming outdated in their traditional form. The risk society is closing the divide, though in a way and from a direction not foreseen by the proponents of turning intelligence analysis into a science.

Coming to Grips with the Uncertainty Divides

As the seconds of terror literally have become more frequent, and as the flood of signals, noise, and interference from media and a growing range of non-state actors increases, the 20th-century positivist concept of intelligence is becoming increasingly unsustainable and counterproductive, "The Truth" is simply not out there, and under no circumstances is it available in time. Intelligence, always under the suspicion of not being able to deliver, will inevitably fall behind, chasing an unattainable goal. Future intelligence has to construct a new role and must do it in order to prevail over the increasing number of competitors providing assessments and advice. One such role, certainly not new but becoming increasingly important in the face of the information explosion, media presence, and the pressure from unfolding events or contingencies, is that of validation, to sort out the relevant from the irrelevant, the corroborated pieces of information from the uncorroborated ones, the flawed assessments from the logically consistent and triangulated. Intelligence in this mode would not deliver products but be a part of a dialogue resting on trust. Intelligence would here need to acquire certain features more associated with academia—a focus of deep knowledge in a range of domains, impartiality, and integrity. Knowledge production for the sake of knowledge production might become a prerequisite to gain and preserve this trust. However, in a competition for talent in which intelligence might find it increasingly hard to attract "the best and the brightest," this will hardly be achievable.

In any case, reforming the expert role would only address one aspect of the challenge. Assessments, however well founded, do not automatically lead to reactions. There is no direct link of causality, sometimes for good reasons but sometimes due to insufficient comprehension, or simply a tendency to ignore issues that are too complex, unwanted, or embedded in boundless uncertainty. The more a dialogue can be established, the less is the perceived risk of dysfunctional politicization and the subsequent need for a safety distance that on the whole would add nothing to overcoming the intelligence-policymaker divide.

There is a need for the development of hybrid organizations, where experts, policymakers, and stakeholders are linked not just through a requirement-dissemination loop but interact on a more continuous and integrated basis. Those hybrid organizations, though, offer both opportunities and negative side effects. On the plus side, the expert-policymaker divide can be addressed more systematically, not least in development of a common terminology and common quantifying of confidence and probability, as done within the Intergovernmental Panel on Climate Change (see Chapter 4).[21] Politicization

---

[21] Intergovernmental Panel on Climate Change: Working Group 1, "Fifth Assessment Report, Chapter 1," pp. 17–19, available at http://www.ipcc.ch/report/ar5/wg1/.

will be a part of such organizations almost by definition, but not necessarily as a drawback. Intelligence has already taken the first steps toward hybrid organizations and cross-disciplinarity in the form of fusion centers, steps taken under pressure from the dilemmas of current intelligence and instant decision making. Breaking up organizational territoriality and professional distance is hardly the final answer, but it does constitute a first step in a more fundamental and inevitable reform of the very concept of intelligence, somewhere along the road approaching or merging with the scientific-stakeholder hybrid organizations addressing similar problems of risk, threat, and uncertainty.

BIBLIOGRAPHY

"Activity Based Intelligence Knowledge Management," USD(I) Concept Paper. Washington, DC: Undersecretary of Defense, Intelligence, Draft, 2011.

Agrell, Wilhelm. "Beyond Cloak and Dagger." In *Clio Goes Spying: Eight Essays on the History of Intelligence*, edited by W. Agrell and B. Huldt. Lund: Lund Studies in International History, 1983.

Agrell, Wilhelm. *The Black Swan and Its Opponents. Early Warning Aspects of the Norway Attacks on July 22, 2011*. Stockholm: National Defence College, Center for Asymetric Threat Studies, 2013.

Agrell, Wilhelm. *Essence of Assessment: Methods and Problems of Intelligence Analysis*. Stockholm: National Defence College, Center for Asymmetric Threat Studies, 2012.

Agrell, Wilhelm. "Selling Big Science: Perceptions of Prospects and Risks in the Public Case for the ESS in Lund." In *In Pursuit of a Promise: Perspectives on the Political Process to Establish the European Spallation Source (ESS) in Lund, Sweden*, edited by Olof Hallonsten. Lund: Arkiv, 2012.

Agrell, Wilhelm. "When Everything Is Intelligence, Nothing Is Intelligence." In *Kent Center Occasional Papers*. Washington, DC: Central Intelligence Agency, 2003.

Aldrich, Richard J. "Policing the Past: Official History, Secrecy and British Intelligence since 1945." *English Historical Review* 119, no. 483 (2004): 922–953.

Aldrich, Richard J. "Regulation by Revelation? Intelligence, the Media and Transparency." In *Spinning Intelligence: Why Intelligence Needs the Media, Why the Media Needs Intelligence*, edited by Robert Dover and Michael S. Goodman. New York: Columbia University Press, 2009.

Allison, Graham T. *Essence of Decision: Explaining the Cuban Missile Crisis*. Boston: Little, Brown, 1971.

Allison, Graham T., and P. Zelikow. *Essence of Decision: Explaining the Cuban Missile Crisis*. New York: Longman, 1999.

Andrus, Calvin. "Toward a Complex Adaptive Intelligence Community: The Wiki and the Blog." *Studies in Intelligence* 49, no. 3 (2005), https://www.cia.gov/library/center-for-the-study-of-intelligence/csi-publications/csi-studies/studies/vol49no3/html_files/Wik_and_%20Blog_7.htm.

"The Apparatgeist Calls." *The Economist* (December 30, 2009).

Armstrong, Fulton. "The CIA and WMDs: The Damning Evidence." *New York Review of Books* (2013), http://www.nybooks.com/articles/archives/2010/aug/19/cia-and-wmds-damning-evidence/?pagination=false.

"As the Enrichment Machines Spin On." *The Economist* (2008), http://www.economist.com/node/10601584.

Bar-Joseph, U. *The Watchman Fell Asleep: The Surprise of Yom Kippur and Its Sources*. Albany: State University of New York Press, 2005.

Barnaby, Frank, and Ronald Huisken. *Arms Uncontrolled*. Cambridge, MA: Harvard University Press, 1975.

Bauer, M. *Resistance to New Technology: Nuclear Power, Information Technology and Biotechnology*. Cambridge: Cambridge University Press, 1994.

Beck, Ulrich. *Risk Society: Towards a New Modernity*. London: Sage, 1992.

Bernard, H. Russell. *Handbook of Methods in Cultural Anthropology*. Walnut Creek, CA: AltaMira Press, 1998.

Bernard, H. Russell. *Research Methods in Anthropology: Qualitative and Quantitative Approaches*. Walnut Creek, CA: AltaMira Press, 2006.

Bernard, H. Russell, Peter Killworth, David Kronenfeld, and Lee Sailer. "The Problem of Informant Accuracy: The Validity of Retrospective Data." *Annual Review of Anthropology* 13 (1984): 495–517.

Bernard, H. Russell, Pertti J. Pelto, Oswald Werner, James Boster, A. Kimball Romney, Allen Johnson, Carol R. Ember, and Alice Kasakoff. "The Construction of Primary Data in Cultural Anthropology." *Current Anthropology* 27, no. 4 (1986): 382–396.

Bernheim, Ernst. *Lehrbuch der historischen Methode und der Geschichtsphilosophie*. Leipzig: Duncker & Humblot, 1908.

Betts, Richard K. *Enemies of Intelligence: Knowledge and Power in American National Security*. New York: Columbia University Press, 2007.

Betts, Richard K. "Two Faces of Intelligence Failure: September 11 and Iraq's Missing WMD." *Political Science Quarterly* 122, no. 4 (2007): 585–606.

Bill Donaldson et al. "MITRE Corporation: Using Social Technologies to Get Connected." *Ivey Business Journal* (January/February 2011).

Bimber, Bruce. *The Politics of Expertise in Congress: The Rise and Fall of the Office of Technology Assessment*. Albany: State University of New York Press, 1996.

Blight, James G., and David A. Welch. *Intelligence and the Cuban Missile Crisis*. London: Frank Cass, 1998.

Bohn, Roger E., and James E. Short. *How Much Information? 2009: Report on American Consumers*. San Diego: Global Information Industry Center, University of California, 2009.

Boin, Arjen, Paul 't Hart, Eric Stern, and Bengt Sundelius. *The Politics of Crisis Management: Public Leadership under Pressure*. Cambridge: Cambridge University Press, 2005.

Buck, Peter. "Adjusting to Military Life: The Social Sciences Go to War 1941–1950." In *Military Enterprise and Technological Change: Perspectives on the American Experience*, edited by Merritt Roe Smith. Cambridge, MA: MIT Press, 1985.

Bumiller, Elisabeth. "Gates on Leaks, Wiki and Otherwise." *New York Times* (November 30, 2010).

Bush, George W. *Decision Points*. New York: Crown, 2011.

Bush, Vannevar. "Science: The Endless Frontier." *Transactions of the Kansas Academy of Science (1903–)* 48, no. 3 (1945): 231–264.

Butler Commission. *Review of Intelligence on Weapons of Mass Destruction*. London: Stationery Office, 2004.

Bynander, Fredrik. *The Rise and Fall of the Submarine Threat: Threat Politics and Submarine Intrusions in Sweden 1980–2002*. Uppsala: Uppsala Acta Universitatis Upsaliensis, 2003.

Calhoun, Mark T. "Clausewitz and Jomini: Contrasting Intellectual Frameworks in Military Theory." *Army History* 80 (2011): 22–37.

Carson, Rachel, *Silent Spring*. Boston: Houghton Mifflin, 1962.

Cassidy, David C. *Uncertainty: The Life and Science of Werner Heisenberg*. New York: W. H. Freeman, 1992.

Center for the Study of Intelligence. *At Cold War's End: US Intelligence on the Soviet Union and Eastern Europe, 1989–1991*. Washington, DC: Central Intelligence Agency, 1999.

Christensen, Tom, Per Lægreid, and Lise H. Rykkja. "How to Cope with a Terrorist Attack?—a Challenge for Political and Administrative Leadership." *In COCOPS Working Paper No 6*, pp. 4, 2012, http://www.cocops.eu/publications/working-papers.

Church Committee Report. *Foreign and Military Intelligence: Book 1: Final Report of the Select Committee to Study Governmental Operations with Respect to Intelligence Activities.* Washington: United States Senate, 1976.

Clarke, D. "Archaeology: The Loss of Innocence." *Archaeology* 47, no. 185 (1973): 6–18.

Cohen, Maurie J. *Risk in the Modern Age: Social Theory, Science and Environmental Decision-Making.* Basingstoke: Palgrave, 2000.

Commission of the European Communities. Directorate-General for Agriculture. *European Community Forest Health Report 1989: Executive Report.* Luxembourg: Office for Official Publications of the European Communities, 1990.

Committee on Behavioral and Social Science Research to Improve Intelligence Analysis for National Security; National Research Council. *Intelligence Analysis for Tomorrow: Advances from the Behavioral and Social Sciences.* Washngton D.C: National Academies Press, 2011.

Darling, Arthur. "The Birth of Central Intelligence," Sherman Kent Center for the Study of Intelligence, www.cia.gov/csi/kent_csi/docs/v10i2a01p_0001.htm.

Davies, Phillip H. J. *Intelligence and Government in Britain and the United States, Volume 1: The Evolution of the U.S. Intelligence Community.* Santa Barbara, CA: Praeger, 2012.

Davies, Phillip H. J. "Intelligence Culture and Intelligence Failure in Britain and the United States." *Cambridge Review of International Affairs* 17, no. 3 (2004): 495–520.

Davies, Phillip H. J., and Kristian C. Gustafson. *Intelligence Elsewhere: Spies and Espionage Outside the Anglosphere*: Washington, DC: Georgetown University Press, 2013.

Dessants, Betty Abrahamsen. "Ambivalent Allies: OSS' USSR Division, the State Department, and the Bureaucracy of Intelligence Analysis, 1941–1945." *Intelligence and National Security* 11, no. 4 (1996): 722–753.

Deutscher, Irwin. "Words and Deeds: Social Science and Social Policy." *Social Problems* 13 (1965): 235.

Dewar, James, Carl H. Builder, William M. Hix, and Morlie H. Levin. *Assumption-Based Planning: A Planning Tool for Very Uncertain Times.* Santa Monica, CA: RAND Corporation, 1993.

"Dommen over Gleditsch Og Wilkes. Fire Kritiske Innlegg." Oslo: Peace Research Institute of Oslo, 1981.

Donaldson, Bill, et al. "MITRE Corporation: Using Social Technologies to Get Connected," *Ivey Business Journal,* June 13, 2011.

Dover, Robert, and Michael S. Goodman, eds. *Spinning Intelligence: Why Intelligence Needs the Media, Why the Media Needs Intelligence.* New York: Columbia University Press, 2009.

Drapeau, Mark, and Linton. *Social Software and National Security: An Initial Net Assessment.* Washington, DC: Center for Technology and National Security Policy, National Defense University, 2009, p. 6.

Drogin, Bob. *Curveball: Spies, Lies, and the Con Man Who Caused a War.* New York: Random House, 2007.

Duelfer, Charles A. *Comprehensive Report of the Special Advisor to the DCI on Iraq's WMD.* Washington, September 30, 2004, https://www.cia.gov/library/reports/general-reports-1/iraq_wmd_2004/

Duelfer, Charles A., and Stephen Benedict Dyson. "Chronic Misperception and International Conflict: The US-Iraq Experience." *International Security* 36, no. 1 (2011): 73–100.

Dulles, Allen. *The Craft of Intelligence.* New York: Harper and Row, 1963.

Eichengreen, Barry, Donghyun Park, and Kwanho Shin. "When Fast Growing Economies Slow Down: International Evidence and Implications for China." *National Bureau of Economic Research R Working Paper* 16919(2011), http://www.nber.org/papers/w16919.

Ekman, Stig. *Den Militära Underrättelsetjänsten. Fem Kriser under Det Kalla Kriget (The Military Intelligence. Five Crises during the Cold War).* Stockholm: Carlsson, 2000.

"Etern Gjennomgang Av Politiets Sikkerhetstjeneste. Rapport Fra Traavikutvalget." Oslo: Justis- Og Beredskapsdepartementet, December 3, 2012.

European Peace Movements and the Future of the Western Alliance. Edited by Walter Laqueur and R. E. Hunter. New Brunswick, NJ: Transaction Books, 1985.

Feinstein, Alvan R. "The 'Chagrin Factor' and Qualitative Decision Analysis." *Archives of Internal Medicine* 145, no. 7 (1985): 1257.

Fingar, Thomas. *Reducing Uncertainty*. Stanford CA: Stanford University Press, 2011.

Fjaestad, Björn. "Why Journalists Report Science as They Do." In *Journalism, Science and Society: Science Communication between News and Public Relations*, edited by Martin W. Bauer and Massimiano Bucchi. New York, Routledge, 2007.

Flores, Robert A., and Joe Markowitz. *Social Software—Alternative Publication @ NGA*. Harper's Ferry, VA: Pherson Associates, 2009.

Flynn, Michael. *Fixing Intel: A Blueprint for Making Intelligence Relevant in Afghanistan*. Washington: Center for a New American Century, 2010.

Foreign Relations of the United States, 1945–1950: Emergence of the Intelligence Establishment. Washington, DC: United States Department of State, 1996.

Forskning eller spionasje. Rapport om straffesaken i Oslo Byrett i mai 1981 (Research or Espionage. Report on the Criminal Trial in Oslo Town Court in May 1981). Oslo: PRIO, 1981.

Frankel, Henry. "The Continental Drift Debate." In *Scientific Controversies: Case Studies in the Resolution and Closure of Disputes in Science and Technology*, edited by Hugo Tristram Engelhardt and Arthur Leonard Caplan. Cambridge: Cambridge University Press, 1987.

Freedman, Lawrence. "The CIA and the Soviet Threat: The Politicization of Estimates, 1966–1977." *Intelligence and National Security* 12, no. 1 (1997): 122–142.

Friedman, Jeffrey A., and Richard Zeckhauser. "Assessing Uncertainty in Intelligence." *Intelligence and National Security* 27, no. 6 (2012): 824–847.

Garthoff, Raymond L. "U.S. Intelligence in the Cuban Missile Crisis." In *Intelligence and the Cuban Missile Crisis*, edited by James G. Blight and David A. Welch. London: Frank Cass, 1998.

Gates, Robert M. "Guarding against Politicization." *Studies in Intelligence* 36, no. 1 (1992): 5–13.

Gazit, Shlomo. "Intelligence Estimates and the Decision-Maker." In *Leaders and Intelligence*, edited by Michael I. Handel. London: Frank Cass, 1989.

George, Roger Zane. "Beyond Analytic Tradecraft." *International Journal of Intelligence and CounterIntelligence* 23, no. 2 (2010): 296–306.

German, Michael, and Jay Stanley. *What's Wrong with Fusion Centers?* New York: American Civil Liberties Union, 2007.

Gibbons, Michael, Camille Limoges, Helga Nowotny, Simon Schwartzman, Peter Scott, and Martin Trow. *The New Production of Knowledge: The Dynamics of Science and Research in Contemporary Societies*. London: Sage, 1994.

Gieryn, Thomas F. "Boundary-Work and the Demarcation of Science from Non-Science: Strains and Interests in Professional Ideologies of Scientists." *American Sociological Review* 48, no. 6 (1983): 781–95.

Gill, David W. J. "Harry Pirie-Gordon: Historical Research, Journalism and Intelligence Gathering in Eastern Mediterranean (1908–18)." *Intelligence and National Security* 21, no. 6 (December 2006): 1045–1059.

Gill, Peter, Stephen Marrin, and Mark Phythian. *Intelligence Theory: Key Questions and Debates*. London: Routledge, 2009.

Gleditsch, Nils Petter. "Freedom of Expression, Freedom of Information, and National Security: The Case of Norway." In *Secrecy and Liberty: National Security, Freedom of Expression and Access to Information*, edited by Sandra Coliver et al. Haag: Nijhoff, 1999, pp. 361–388.

Gleditsch, Nils Petter, Sverre Lodgaard, Owen Wilkes, and Ingvar Botnen. *Norge i atomstrategien. Atompolitikk, alliansepolitikk, basepolitikk (Norway in the Nuclear Strategy. Nuclear Policy, Alliance Policy, Base Policy)*. Oslo: PAX, 1978.

Gleditsch, Nils Petter, and Owen Wilkes. *Forskning om etterretning eller etterretning som forskning*. Oslo: Peace Research Institute of Oslo, 1979.

Gleditsch, Nils Petter, and Owen Wilkes. *Intelligence Installations in Norway: Their Number, Location, Function, and Legality*. Oslo: International Peace Research Institute, 1979.

Gleditsch, Nils Petter, and Høgetveit, Einar. "Freedom of Information and National Security. A Comparative Study of Norway and United States." *Journal of Peace Research* 2, no. 1 (1984): 17–45.

Göpfert, Winfred. "The Strength of PR and Weakness of Science Journalism." In *Journalism, Science and Society: Science Communication between News and Public Relations*, edited by Martin W. Bauer and Massimiano Bucchi. New York: Routledge, 2007.

Hadley, Stephen. "Press Briefing by National Security Advisor Stephen Hadley." Washington, D.C., December 3, 2007, http://iipdigital.usembassy.gov/st/english/texttrans/2007/12/20071203191509xjsnommis0.1537287.html#axzz38FO0ulH6.

Harris, Shane. "The Other About-Face on Iran, *National Journal*, December 14, 2007, accessed February 21, 2013, http://shaneharris.com/magazinestories/other-about-face-on-iran/

Hartcup, Guy. *The War of Invention: Scientific Developments, 1914–18*. London: Brassey's Defence Publishers, 1988.

Hellström, Tomas, and Merle Jacob. *Policy Uncertainty and Risk: Conceptual Developments and Approaches*. Dordrecht: Kluwer, 2001.

Herbert, Matthew. "The Intelligence Analyst as Epistemologist." *International Journal of Intelligence and CounterIntelligence* 19, no. 4 (2006): 666–684.

Herman, Michael. *Intelligence Power in Peace and War*: Cambridge: Cambridge University Press, 1996.

Heuer, Richards J. *Psychology of Intelligence Analysis*. Washington, DC: Center for the Study of Intelligence, Central Intelligence Agency, 1999.

Hilsman, Roger. "Intelligence and Policy-Making in Foreign Affairs." *World Politics* 5 (1952): 1–45.

Hinsley, Francis H., and Alan Stripp. *Codebreakers: The Inside Story of Bletchley Park*. Oxford: Oxford University Press, 2001.

Horowitz, Louis, ed. *The Rise and Fall of Project Camelot: Studies in the Relationship between Social Science and Practical Politics*. Cambridge, MA: MIT Press, 1967.

"A Historical Review of Studies of the Intelligence Community for the Commission on the Organization of the Government for the Conduct of Foreign Policy." (document TS-206439-74). Washington, D.C.: Commission on the Organization of Government for the Conduct of Foreign Policy, 1974.

Hogenboom, Joris, Arthur P. J. Mol, and Gert Spaargen. "Dealing with Environmental Risk in Reflexive Modernity." In *Risk in the Modern Age: Social Theory, Science and Environmental Decision-Making*, edited by Maurie J. Cohen. Basingstoke: Palgrave, 2001.

Hoover, Calvin B. "Final Report (No Date) Rg 226 (OSS), Entry 210 Box 436." College Park, MD: National Archives.

Howard, Alex. "Defining Gov 2.0 and Open Government." January 5, 2011, http://gov20.gov-fresh.com/social-media-fastfwd-defining-gov-2-0-and-open-government-in-2011/.

Hulme, Mike, and Martin Mahony. "Climate Change: What Do We Know about the IPCC?" *Progress in Physical Geography* 34, no. 5 (2010): 705–718.

Hulnick, Arthur. "What's Wrong with the Intelligence Cycle." In *Strategic Intelligence* edited by Loch Johnson. Westport CT: Greenwood, 2007.

Hutton, B., and Great Britain, Parliament, House of Commons. *Report of the Inquiry into the Circumstances Surrounding the Death of Dr. David Kelly C.M.G.* London: Stationery Office, 2004.

Imai, Masaaki. *Kaizen: The Key to Japan's Competitive Success*. New York: McGraw-Hill, 1986.

Intentions and Capabilities. Estimates on Soviet Strategic Forces, 1950–1983. Edited by Donald P. Steury. Washington, DC: Center for the Study of Intelligence, Central Intelligence Agency, 1996.

InterAcademy Council. *Climate Change Assessments: Review of the Processes and Procedures of the IPCC*. Amsterdam: InterAcademy Council, 2010.

Intergovernmental Panel on Climate Change. *Climate Change 2013—the Physical Science Basis: Working Group I Contribution to the Fifth Assessment Report of the IPCC (Preliminary Report)*. Cambridge: Cambridge University Press, 2013.

Intergovernmental Panel on Climate Change. *Principles Governing IPCC Work*. Vienna: IPCC, 1998.

Intergovernmental Panel on Climate Change: Working Group 1. *Fifth Assessment Report*, http://www.ipcc.ch/report/ar5/wg1/.

Interview by author (Treverton) with Stephen Hadley. 2012.

Janis, Irving L. *Groupthink: Psychological Studies of Policy Decisions and Fiascoes.* Boston: Houghton Mifflin, 1972.

Jervis, Robert. *Perceptions and Misperceptions in International Politics.* Princeton, NJ: Princeton University Press, 1976.

Jervis, Robert. *Why Intelligence Fails: Lessons from the Iranian Revolution and the Iraq War.* Ithaca, NY: Cornell University Press, 2010.

Johnson, Loch. "The CIA and the Media." *Intelligence and National Security* 1 (2) May 1986: 143–169.

Johnston, Rob. "Analytic Culture in the US Intelligence Community: An Ethnographic Study." Washington, DC: Center for Study of Intelligence,Central Intelligence Agency, 2005.

Joint Intelligence Committee. *Iraq's Weapons of Mass Destruction: The Assessment of the British Government.* London: Stationery Office, 2002.

Jomini, Antoine Henri Baron de. *The Art of War.* Translated by G. H. Mendell and W. P. Craighill. Mineola, NY: Dover, 2007.

Jones, R. V. *Instruments and Experiences: Papers on Measurement and Instrument Design.* Hoboken, NJ: Wiley, 1988.

Jones, R. V. *Most Secret War: British Scientific Intelligence 1939–45.* London: Coronet Books, 1979.

Jones, R. V. *Reflections on Intelligence.* London: Mandarin, 1990.

Joravsky, David. *The Lysenko Affair.* Cambridge, MA: Harvard University Press, 1970.

Jungk, Robert. *Brighter Than a Thousand Suns: A Personal History of the Atomic Scientists.* Harmondsworth: Penguin Books, 1982.

Kam, Ephraim. *Surprise Attack: The Victim's Perspective.* Cambridge: Cambridge University Press, 1988.

Kätz, Barry. *Foreign Intelligence: Research and Analysis in the Office of Strategic Services, 1942–1945.* Cambridge, MA: Harvard University Press, 1989.

Keegan, J. *Intelligence in War: Knowledge of the Enemy from Napoleon to Al-Qaeda.* New York: Knopf Doubleday, 2003.

Kendall, Willmoore. "The Function of Intelligence." *World Politics* 1, no. 4 (1949): 542–552.

Kent, Sherman. "A Crucial Estimate Relived." *Studies in Intelligence* 8, no. 2 (1964): 1–18.

Kent, Sherman. "The Need for an Intelligence Literature." *Studies in Intelligence* 1, no. 1 (1955): 1–8.

Kent, Sherman. *Strategic Intelligence for American World Policy.* Princeton, NJ: Princeton University Press, 1949.

Kent, Sherman. "Words of Estimative Probability." *Studies in Intelligence* 8, no. 4 (Fall 1964): 49–65.

Kent, Sherman. *Writing History.* New York: F.S. Crofts, 1941.

Kerbel, Josh, and Anthony Olcott. "Synthesizing with Clients, Not Analyzing for Customers." *Studies in Intelligence* 54, no. 4 (2010): 11–27.

Kissinger, H. *Years of Renewal: The Concluding Volume of His Classic Memoirs.* London: Simon and Schuster, 2012.

Kissinger, H. *Years of Upheaval.* London: Weidenfeld and Nicolson, 1982.

Kissinger, H. *Years of Upheaval: The Second Volume of His Classic Memoirs.* London: Simon and Schuster, 2011.

Klitgaard, Robert. "Policy Analysis and Evaluation 2.0." Unpublished paper, 2012.

Knorr, K. E. *Foreign Intelligence and the Social Sciences.* Princeton, NJ: Center of International Studies, Woodrow Wilson School of Public and International Affairs, Princeton University, 1964.

Krimsky, Sheldon. *Science in the Private Interest: Has the Lure of Profits Corrupted Biomedical Research?* Lanham, MD: Rowman and Littlefield, 2003.

Kristin Quinn. "A Better Toolbox." *TrajectoryMagazine.com*, 2012.

Kuhn, Thomas S. *The Structure of Scientific Revolutions.* Chicago: University of Chicago Press, 1962.

LaFollette, Marcel C. *Stealing into Print: Fraud, Plagiarism, and Misconduct in Scientific Publishing.* Berkeley: University of California Press, 1992.

Landsberger, Henry A. *Hawthorne Revisited: Management and the Worker, Its Critics, and Developments in Human Relations in Industry.* Ithaca: New York State School of Industrial and Labor Relations, 1958.

Laudani, Raffaele, ed. *Secret Reports on Nazi Germany: The Frankfurt School Contribution to the War Efforts. Franz Neumann, Herbert Marcuse, Otto Kirchheimer.* Princeton, NJ: Princeton University Press, 2013.

Laqueur, Walter. "The Question of Judgment: Intelligence and Medicine." *Journal of Contemporary History* 18, no. 4 (1983): 533–548.

Laqueur, Walter. *A World of Secrets: The Uses and Limits of Intelligence.* London: Weidenfeld and Nicolson, 1985.

Lefebvre, Stéphane. "A Look at Intelligence Analysis." *International Journal of Intelligence and CounterIntelligence* 17, no. 2 (2004): 231–264.

Lehner, Paul, Avra Michelson, and Leonard Adelman. *Measuring the Forecast Accuracy of Intelligence Products.* Washington, DC: The MITRE Corporation, 2010.

Lennon, Michael, and Gary Berg-Cross. "Toward a High Performing Open Government." *The Public Manager* 39, no. 10 (Winter 2010). http://www.astd.org/Publications/Magazines/The-Public-Manager/Archives/2010/10/Toward-a-High-Performing-Open-Government?.

"Lessons Learned from Intelligence Successes, 1950–2008 (U)." Kent School Occasional Paper, Washington, D.C.: Central Intelligence Agency, io, 2010.

Liberatore, Angela. *The Management of Uncertainty: Learning from Chernobyl.* Amsterdam: Gordon and Breach, 1999.

Lippmann, Walter. *Public Opinion.* New York: Harcourt, Brace, 1922.

Litwak, Robert S. "Living with Ambiguity: Nuclear Deals with Iran and North Korea." *Survival* 50, no. 1 (2008): 91–118.

Lock, Stephan. *Fraud and Misconduct in Medical Research.* Edited by Frank Wells. London: BMJ Publishing Group, 1993.

Lock, Stephan, and Frank Wells. *Fraud and Misconduct: In Biomedical Research.* Edited by Michael Farthing. London: BMJ Books, 2001.

Lowenthal, Mark M. "A Disputation on Intelligence Reform and Analysis: My 18 Theses." *International Journal of Intelligence and CounterIntelligence* 26, no. 1 (2013): 31–37.

Lucas, R.E., Jr. "Econometric Policy Evaluation: A Critique." In *The Phillips Curve and Labor Markets, Carnegie-Rochester C Carnegie-Rochester Conference on Public Policy,* edited by K. Brunner and A.H. Meltzer, Vol. 1. Amsterdam: North Holland, 1976, 19–46.

Lundgren, Lars J. *Acid Rain on the Agenda: A Picture of a Chain of Events in Sweden, 1966–1968.* Lund: Lund University Press, 1998.

MacEachin, D. J. *Predicting the Soviet Invasion of Afghanistan: The Intelligence Community's Record.* Washington, DC: Center for the Study of Intelligence, Central Intelligence Agency, 2002.

Marks, J. H. "Interrogational Neuroimaging in Counterterrorism: A No-Brainer or a Human Rights Hazard." *American Journal of Law & Medicine* 33 (2007): 483.

Marquardt-Bigman, Petra. "The Research and Analysis Branch of the Office of Strategic Services in the Debate of US Policies towards Germany, 1943–46." *Intelligence and National Security* 12, no. 2 (1997): 91–100.

Marrin, Stephen. "Best Analytic Practices from Non-Intelligence Sectors." edited by Analytics Institute, 2011.

Marrin, Stephen. *Improving Intelligence Analysis: Bridging the Gap between Scholarship and Practice.* London: Routledge, 2011.

Marrin, Stephen. "Preventing Intelligence Failures by Learning from the Past." *International Journal of Intelligence and CounterIntelligence* 17, no. 4 (2004).

Marrin, Stephen. "Intelligence Analysis: Turning a Craft into a Profession." *International Conference on Intelligence Analysis, Proceedings of the First International Conference on Intelligence Analysis,* McLean, VA: MITRE, May, 2005.

Marrin, Stephen, and Jonathan D. Clemente. "Improving Intelligence Analysis by Looking to the Medical Profession." *International Journal of Intelligence and CounterIntelligence* 18, no. 4 (2005): 707–729.

Marrin, Stephen, and Jonathan D. Clemente. "Modeling an Intelligence Analysis Profession on Medicine 1." *International Journal of Intelligence and CounterIntelligence* 19, no. 4 (2006): 642–665.

Masse, Todd, Siobhan O'Neil, and John Rollins. *Fusion Centers: Issues and Options for Congress.* Washington, DC: Congressional Research Service, 2007.

McAuliffe, Mary S. *CIA Documents on the Cuban Missile Crisis, 1962.* Washington, DC: Central Intelligence Agency, 1992.

McConnell, Michael. "Intelligence Community Directive Number 203." Washington, DC: Office of the Director of National Intelligence, 2007.

McConnell, Michael. "Memorandum: Guidance on Declassification of National Intelligence Estimate Key Judgments." Washington, DC: Office of the Director of National Intelligence, 2007.

McDermott, Rose. "Experimental Intelligence." *Intelligence and National Security* 26, no. 1 (2011): 82–98.

McFate, Montgomery. "Anthropology and Counterinsurgency: The Strange Story of Their Curious Relationship." *Military Review* 85, no. 2 (2005): 24–38.

McFate, Montgomery, and Steve Fondacaro. "Cultural Knowledge and Common Sense." *Anthropology Today* 24, no. 1 (2008): 27.

McFate, Montgomery, and Steve Fondacaro. "Reflections on the Human Terrain System during the First 4 Years." *Prisms* 2, no. 4 (2011): 63–82.

McKay, C. G., and Bengt Beckman. *Swedish Signal Intelligence: 1900–1945.* London: Frank Cass, 2003.

McMullin, Ernan. "Scientific Controversy and Its Termination." In *Scientific Controversies: Case Studies in the Resolution and Closure of Disputes in Science and Technology,* edited by Hugo Tristram Engelhardt and Arthur Leonard Caplan. Cambridge: Cambridge University Press, 1987.

Medvedev, Z. A., and T. D. Lysenko. *The Rise and Fall of T.D. Lysenko.* New York: Columbia University Press, 1969.

Mihelic, Matthew. "Generalist Function in Intelligence Analysis." International Conference on Intelligence Analysis, 2005.

Militärpolitik och stridskrafter—läge och tendenser 1975, Överbefälhavaren Specialorientering (Military Policy and Force Postures—Situation and Tendencies 1975, Supreme Commander Special Report 1975-01-14)." Stockolm: Försvarsstaben, 1975.

Miller, David J., and Michael Hersen. *Research Fraud in the Behavioral and Biomedical Sciences.* New York: John Wiley, 1992.

Miller, F. P., A. F. Vandome, and M. B. John. *Jacobellis v. Ohio.* VDM Publishing, 2011.

Morrison, John N. L. "British Intelligence Failures in Iraq." *Intelligence and National Security* 26, no. 4 (2011): 509–520.

Moseholm, Lars, Bent Andersen, and Ib Johnsen. *Acid Deposition and Novel Forest Decline in Central and Northern Europe: Assessment of Available Information and Appraisal of the Scandinavian Situation: Final Report, December 1986.* Copenhagen: Nordic Council of Ministers, 1988.

National Commission on Terrorist Attacks upon the United States. *The 9/11 Commission Report: Final Report of the National Commission on Terrorist Attacks upon the United States.* Washington, DC: US Government Printing Office, 2004.

National Intelligence Council. "How the Intelligence Community Arrived at the Judgments in the October 2002 NIE on Iraq's WMD Programs." Washington, DC: National Intelligence Council, 2004.

National Intelligence Council. *Iran: Nuclear Intentions and Capabilities.* National Intelligence Estimate, 2007. The Key Judgments were declassified and are avilable at http://www.counterwmd.gov/files/20071203_release.pdf.

National Intelligence Council. *Yugoslavia Transformed.* National Intelligence Estimate (15-90), 1990. The entire estimate was declassified and is available at at http://www.foia.cia.gov/search-results?search_api_views_fulltext=Yugoslavia&field_collection=11.

Nelkin, Dorothy. *Selling Science: How the Press Covers Science and Technology.* New York: Freeman, 1995.

Neustadt, Richard E., and Harvey V. Fineberg. "The Swine Flu Affair: Decision-Making on a Slippery Disease." In *Kennedy School of Government Case C14-80-316*. Cambridge, MA: Harvard University.

Nickerson, Raymond S. "Confirmation Bias: A Ubiquitous Phenomenon in Many Guises." *Review of General Psychology* 2, no. 2 (1998): 175–220.

Nomination of Robert M. Gates, Hearings before the Select Committee on Intelligence of the United States Senate. Washington, DC: US Government Printing Office, 1992, 510–511.

Northern Securities Co. v. United States, 193 U.S. 197, 400–411 (1904).

Nye, Joseph S. Jr. "Peering into the Future." *Foreign Affairs* (July/August 1994): 82–93.

O'Reilly, Jessica, Naomi Oreskes, and Michael Oppenheimer. "The Rapid Disintegration of Projections: The West Antarctic Ice Sheet and the Intergovernmental Panel on Climate Change." *Social Studies of Science* 42, no. 5 (2012): 709–731.

Olcott, Anthony. "Institutions and Information: The Challenge of the Six Vs." Washington, DC: Institute for the Study of Diplomacy, Georgetown University., April 2010.

Olcott, Anthony. "Revisiting the Legacy: Sherman Kent, Willmoore Kendall, and George Pettee—Strategic Intelligence in the Digital Age." *Studies in Intelligence* 53, no. 2 (2009): 21–32.

Olcott, Anthony. "Stop Collecting—Start Searching (Learning the Lessons of Competitor Intelligence)." Unpublished paper.

Palm, Thede. *Några Studier Till T-Kontorets Historia*. Vol. 21, Kungl. Samgfundet for Utgivande Av Handskrifter Rorande, Stockholm: Kungl. Samfundet för utgivande av handskrifter rörande Skandinaviens historia, Vol. 21, 1999.

Pascovich, Eyal. "Intelligence Assessment Regarding Social Developments: The Israeli Experience." *International Journal of Intelligence and CounterIntelligence* 26, no. 1 (2013): 84–114.

Patrick Kelly. "The Methodology of Ernst Bernheim and Legal History." http://www.scribd.com/doc/20021802/The-Methodology-of-Ern ...

Patton, M. Q. "Use as a Criterion of Quality in Evaluation." In *Visions of Quality: How Evaluators Define, Understand and Represent Program Quality: Advances in Program Evaluation*, edited by A. Benson, C. Lloyd, and D. M. Hinn. Kidlington, UK: Elsevier Science, 2001.

Persson, Gudrun. *Fusion Centres—Lessons Learned. A Study of Coordination Functions for Intelligence and Security Services* (Stockholm: Center for Asymmetric Threat Studies, Swedish National Defence College, 2013).

Peterson, Steven W. *US Intelligence Support to Decision Making*. Weatherhead Center for International Affairs, Harvard University, July 1, 2009.

Pettee, George *The Future of American Secret Intelligence*. Washington: Infantry Journal Press, 1946.

Phythian, Mark. "Policing Uncertainty: Intelligence, Security and Risk." *Intelligence and National Security* 27, no. 2 (2012): 187–205.

Platt, W. *Strategic Intelligence Production: Basic Principles*. New York: F. A. Praeger, 1957.

Popper, Karl Raimund. *Conjectures and Refutations: The Growth of Scientific Knowledge*. London: Routledge and Kegan Paul, 1969.

Popper, Karl Raimund. *The Open Society and Its Enemies*, Vol.1: *The Spell of Plato*. London: Routledge and Kegan Paul, 1945.

Popper, Karl Raimund. *Open Society and Its Enemies*, Vol. 2: *The High Tide of Prophecy; Hegel, Marx and the Aftermath*. London: Routledge and Kegan Paul, 1947.

Popper, Karl Raimund. *The Poverty of Historicism*. London: Routledge and Kegan Paul, 1960.

Prados, John. *The Soviet Estimate: U.S. Intelligence Analysis & Russian Military Strength*. New York: Dial Press, 1982.

Pritchard, Matthew C., and Michael S. Goodman. "Intelligence: The Loss of Innocence." *International Journal of Intelligence and CounterIntelligence* 22, no. 1 (2008): 147–164.

"Probing the Implications of Changing the Outputs of Intelligence: A Report of the 2011 Analyst-IC Associate Teams Program." In *Studies in Intelligence* 56, no. 1, Extracts, 2012, 1–11.

Protest and Survive. Edited by Dan Smith and Edward Palmer Thompson. Harmondsworth: Penguin, 1980.

Prunckun, Henry. "The Intelligence Analyst as Social Scientist: A Comparison of Research Methods." *Police Studies* 19, no. 3 (1996): 70–72.

"Regiosterstudie Med Focus På Neurologiska Och Immunrelaterade Sjukdomar Efter Vaccination Med Pandemrix, Medical Production Agency 26 March, 2013."

Reiss, Albert J. Jr. "The Institutionalization of Risk." In *Organizations, Uncertainties, and Risk,* edited by James F. Short Jr., and Lee Clarke. Boulder, CO: Westview Press, 1992.

Remröd, Jan. *Forest in Danger.* Djursholm: Swedish Forestry Association, 1985.

Respectfully Quoted: A Dictionary of Quotations. Edited by Suzy Platt. Washington, DC: Library of Congress, 1989.

A Review of the Intelligence Community. Washington, DC: Office of Management and Budget, 1975.

Rhodes, Richard. *The Making of the Atomic Bomb.* New York: Simon and Schuster, 1986.

Richelson, Jeffery T. *A Century of Spies: Intelligence in the Twentieth Century:* New York: Oxford University Press, 1995.

The Rise and Fall of Project Camelot. Studies in the Relationship between Social Science and Pratical Politics. Edited by Irving Louis Horowitz. Cambridge, MA: MIT Press, 1967.

Riste, Olav. "The Intelligence-Policy Maker Relationship and the Politicization of Intelligence." In *National Intelligence Systems,* edited by Gregory F. Treverton and Wilhelm Agrell. Cambridge: Cambridge University Press, 2009.

Ritchey, Tom. *Structuring Social Messes with Morphological Analysis.* Stockholm, Swedish Morphological Society, 2007.

Robinson, Mike. "A Failure of Intelligence." In TV-program *BBC Panorama,* 09/07/2004, London: BBC, 2004.

Roese, Neal J., and Kathleen D. Vohs. "Hindsight Bias." *Perspectives on Psychological Science* 7, no. 5 (2012): 411–426.

Rohde, David. "Army Enlists Anthropology in War Zones." *New York Times,* October 5, 2007, http://www.nytimes.com/2007/10/05/world/asia/05afghan.html?incamp=article_popular_4&pagewanted=all&_r=0.

Roll-Hansen, Nils. *Ideological Obstacles to Scientific Advice in Politics: The Case of "Forest Death" from "Acid Rain."* Makt-og demokratiutredningens skriftserie vol. 48. Oslo: Makt- og demokratiutredningen, 2002.

Ronge, Max. *Kriegs-Und Industrie-Spionage: Zwölf Jahre Kundschaftsdienst.* Wien: Amalthea-Verl, 1930.

Rose, Paul Lawrence. *Heisenberg and the Nazi Atomic Bomb Project: A Study in German Culture.* Berkeley: University of California Press, 1998.

Round Two. *The Norwegian Supreme Court vs. Gleditsch & Wilkes,* Oslo: PRIO, 1982.

Rowen, Henry S. "When Will the Chinese People Be Free?" *Journal of Democracy* 18, no. 3 (2007): 38–52.

Rowen, Henry S. "Will China's High Speed Ride Get Derailed?" Unpublished paper, 2011.

Sackett, David L. "Evidence-Based Medicine." *Seminars in Perinatology* 21, no. 1 (1997): 3–5.

Salomon, Jean Jacques. *Science and Politics.* Cambridge, MA: MIT Press, 1973.

Sanchez, Phillip L. *Increasing Information Sharing among Independent Police Departments.* Monterrey, CA: Naval Postgraduate School, 2009.

Sanger, David E. *The Inheritance: The World Obama Confronts and the Challenges to American Power.* New York: Random House, 2010.

Schmidle, Nicholas. "Are Kosovo's Leaders Guilty of War Crimes?" *The New Yorker,* May 6, 2013.

Schmidt-Eenboom, Erich. *Der Schattenkrieger: Klaus Kinkel Und Der BND.* Dusseldorf: ECON, 1995.

Schweber, Silvan S. *In the Shadow of the Bomb: Oppenheimer, Bethe, and the Moral Responsibility of the Scientist.* Princeton, NJ: Princeton University Press, 2000.

"Secrets Not Worth Keeping: The Courts and Classified Information." *Washington Post,* February 15, 1989.

Senate Committee on Homeland Security and Governmental Affairs, Report of the Permanent Subcommittee on Investigations. *Federal Support for and Involvement in State and Local Fusion Centers.* 2. Washington, DC: 2012.

Shackley, Simon. "The Intergovernmental Panal on Climate Change: Consensual Knowledge and Global Politics." *Global Environmental Change* 7 (1997): 77–79.

Sherman, Chris, and Gary Price. *The Invisible Web: Uncovering Information Sources Search Engines Can't See*. Meford, NJ: Information Today, 2001.

Shimshoni, Jonathan. "Technology, Military Advantage, and World War I: A Case for Military Entrepreneurship." *International Security* 15, no. 3 (1990): 187–215.

Shulman, S. W. *The Internet Still Might (but Probably Won't) Change Everything: Stakeholder Views on the Future of Electronic Rulemaking*. Pittsburgh: University of Pittsburgh, University Center for Social and Urban Research, 2004.

Silberman, L. H., and C. S. Robb. *The Commission on the Intelligence Capabilities of the United States Regarding Weapons of Mass Destruction: Report to the President of the United States*. Washington D.C: Executive Agency Publications, 2005.

Silver, N. *The Signal and the Noise: Why So Many Predictions Fail—but Some Don't*. New York: Penguin, 2012.

Sluka, Jeffrey A. "Curiouser and Curiouser: Montgomery Mcfate's Strange Interpretation of the Relationship between Anthropology and Counterinsurgency." *PoLar: Political and Legal Anthropology Review* 33, no. 1 (2010): 99–115.

Smith, Aaron Whitman. *Government Online: The Internet Gives Citizens New Paths to Government Services and Information*. Washington, D.C.: Pew Internet & American Life Project, 2010.

Snowden, David. "Complex Acts of Knowing: Paradox and Descriptive Self-Awareness." *Journal of Knowledge Management* 6, no. 2 (2002): 100–111.

Solow, R. M. "Forty Years of Social Policy and Policy Research." Inaugural Robert Solow Lecture. Washington, DC: Urban Institute, 2008.

Spink, Amanda, Bernard J. Jansen, Vinish Kathuria, and Sherry Koshman. "Overlap among Major Web Search Engines." In *Third International Conference on Information Technology: New Generations*. 2006. Los Alamitos, CA, Institute of Electrical and Electronics Engineers.

Steinbruner, John D., Paul C. Stern, and Jo L. Husbands. *Climate and Social Stress: Implications for Security Analysis*. National Academies Press, 2013.

Stewart, Daniel M., and Robert G. Morris. "A New Era of Policing? An Examination of Texas Police Chiefs' Perceptions of Homeland Security." *Criminal Justice Policy Review* 20, no. 3 (2009): 290–309.

Stoltzenberg, Dietrich. *Fritz Haber. Chemiker, Nobelpreisträger, Deutscher Jude*. Weinhem: VCH, 1994.

Suits, C. G, George R. Harrison, and Louis Jordan. *Science in World War II. Applied Physics, Electronics, Optics, Metallurgy*. Boston: Little Brown, 1948.

"Swine Flu (E): Summary Case, Kennedy School of Government Case C14-83-527." Cambridge, MA, Harvard University.

Tennison, Michael N., and Jonathan D. Moreno. "Neuroscience, Ethics, and National Security: The State of the Art." *PLoS biology* 10, no. 3 (2012), http://www.plosbiology.org/article/info%3Adoi%2F10.1371%2Fjournal.pbio.1001289.

Tetlock, P. *Expert Political Judgment: How Good Is It? How Can We Know?* Princeton, NJ: Princeton University Press, 2005.

The 22 July Commission. "Findings of the 22 July Commission." Oslo, 2012.

The Oslo Rabbit Trial. A Record of the "National Security Trial" against Owen Wilkes and Nils Petter Gleditsch in the Oslo Town Court, May 1981, Oslo: PRIO, 1981.

Tol, Richard S. J. "Regulating Knowledge Monopolies: The Case of the IPCC." *Climatic Change* 108 (2011): 827–839.

Treverton, Gregory F. "Assessing Threats to the U.S." In *Great Decisions 2013*, pp. 101ff. New York: Foreign Policy Association, 2012.

Treverton, Gregory F. "Estimating beyond the Cold War." *Defense Intelligence Journal* 3, no. 2 (1994): 5–20.

Treverton, Gregory F. "The "First Callers": The President's Daily Brief (PDB) across Three Administrations." Center for the Study of Intelligence, forthcoming.

Treverton, Gregory F. "Intelligence Analysis: Between 'Politicization' and Irrelevance." In *Analyzing Intelligence: Origins, Obstacles, and Innovations,* edited by Roger Z. George and James B. Bruce. Washington, DC: Georgetown University Press, 2008.

Treverton, Gregory F. *Intelligence for an Age of Terror.* New York: Cambridge University Press, 2009.

Treverton, Gregory F. "Learning from Recent Best Practices and Spectacular Flops." In *A Handbook of the Psychology of Intelligence Analysis,* edited by Richard L. Rees. Washinton, D.C.: Central Intelligence Agency, 2007.

Treverton, Gregory F. *New Tools for Collaboration: The Experience of the U.S. Intelligence Community,* forthcoming from IBM Business of Government.

Treverton, Gregory F. *Reshaping National Intelligence for an Age of Information.* New York: Cambridge University Press, 2003.

Treverton, Gregory F. "Risks and Riddles: The Soviet Union Was a Puzzle. Al Qaeda Is a Mystery. Why We Need to Know the Difference." *Smithsonian* (June 2007), http://www.smithsonianmag.com/people-places/risks-and-riddles-154744750/.

Treverton, Gregory F. "What Should We Expect of Our Spies." *Prospect* (June 2011), http://www.prospectmagazine.co.uk/features/what-should-expect-spies-intelligence-mi5-cia-soviet-union-saddam-wmd-arab-spring.

Treverton, Gregory F., and Wilhelm Agrell. *National Intelligence Systems.* New York: Cambridge University Press, 2009.

Treverton, Gregory F., and C. Bryan Gabbard. *Assessing the Tradecraft of Intelligence Analysis.* Santa Monica, CA: RAND, 2008.

Treverton, Gregory F., and Lisa Klautzer. *Frameworks for Domestic Intelligence and Threat Assessment: International Comparisons.* Stockholm: Center for Asymmetric Threat Studies, Swedish National Defence College, 2010.

Tunlid, Anna. "Ett konfliktfyllt fält: förtroende och trovärdighet inom miljöforskningen." in *Forskningens gråzoner—tillrättaläggande, anpassning och marknadsföring i kunskaps-produktion,* edited by Wilhelm Agrell. Stockholm: Carlssons, 2007.

Ulrich, B. "Die Wälder in Mitteneuropa. Messergegnisse Ihrer Umweltbelastung, Theorie Ihre Gefärdung, Prognosen Ihre Entwicklung." *Allgemeine Forstzeitschift* 35, no. 44 (1980): 1198–1202.

United States Congress—Select Committee on Intelligence. *Report on the U.S. Intelligence Community's Prewar Intelligence Assessments on Iraq.* Washington, DC: United States Senate, 2004.

United States Congress. *Intelligence Reform and Terrorism Prevention Act of 2004 (to Accompany S. 2845).* Washington, DC: United States Senate, 2004.

van der Heide, Liesbeth. "Cherry-Picked Intelligence. The Weapons of Mass Destruction Dispositive as a Legitimation for National Security in the Post 9/11 Age." *Historical Social Research* 38, no. 1 (2013): 286–307.

Von Clausewitz, C. *On War.* Translated by Michael Howard and P. Paret. Princeton, NJ: Princeton University Press, 1976.

Wark, Wesley. *The Ultimate Enemy: British Intelligence and Nazi Germany, 1933–1939.* London: I. B. Tauris, 1985.

Watts, Duncan J. *Everything Is Obvious: Once You Know the Answer.* New York: Crown, 2011.

Weart, Spencer R. *The Discovery of Global Warming.* Cambridge, MA: Harvard University Press, 2003.

Weart, Spencer R. "Nuclear Fear: A History and an Experiment." In *Scientific Controversies: Case Studies in the Resolution and Closure of Disputes in Science and Technology,* edited by Hugo Tristram Engelhardt and Arthur Leonard Caplan. Cambridge: Cambridge University Press, 1987.

Weiss, C. H. "The Many Meanings of Research Utilization." *Public Administration Review* 39, no. 5 (1979): 426–431.

West, Nigel, and Oleg Tsarev. *The Crown Jewels.* London: HarperCollins, 1999.

Wilkes, Owen, and Nils Petter Gleditsch. *Onkel Sams kaniner. Teknisk etterretning i Norge* (Uncle Sam's Rabbits. Technical Intelligence in Norway). Oslo: PAX, 1981.

Wilkes, Owen, and Gleditsch, Nils Petter. "Research on Intelligence and Intelligence as Research." In *Elements of World Instability: Armaments, Communication, Food, International Division of Labor, Proceedings of the Eighth International Peace Research Association Conference*, edited by Egbert Jahn and Yoshikazu Sakamoto. Frankfurt/New York: Campus, 1981.

Willis, Henry H., Genevieve Lester, and Gregory F. Treverton. "Information Sharing for Infrastructure Risk Management: Barriers and Solutions." *Intelligence and National Security* 24, no. 3 (2009): 339–365.

Wohlstetter, Roberta. *Pearl Harbor: Warning and Decision*. Palo Alto, CA: Stanford University Press, 1962.

Wolf, Charles, K. C. Yeh, Benjamin Zycher, Nicholas Eberstadt, and Sungho Lee. *Fault Lines in China's Economic Terrain*. MR-1686-NA/SRF ed. Santa Monica, CA: RAND, 2003.

Zehr, Stephen. "Comparative Boundary Work: US Acid Rain and Global Climate Change Policy Deliberations." *Science and Public Policy* 32, no. 6 (2005): 445–456.

Zolling, Hermann, and Heinz Höhne. *Pullach Intern: General Gehlen Und Die Geschichte Des Bundesnachrichtendienstes*. Hamburg: Hoffmann und Campe, 1971.

Zuckerman, Solly. *Beyond the Ivory Tower: The Frontiers of Public and Private Science*. New York: Taplinger, 1971.

INDEX

Figures and tables are indicated by f and t following the page number.